ENVIRONMENTAL LEGACY OF THE MANHATTAN PROJECT

Charleston, SC
www.PalmettoPublishing.com

Environmental Legacy of the Manhattan Project
Copyright © 2023 by Ed Struzeski, Jr.

Paperback ISBN: 979-8-8229-1501-5

ENVIRONMENTAL LEGACY OF THE MANHATTAN PROJECT

Ed Struzeski, Jr.

My sincere appreciation is given to the dedicated environmentalists
whose stories in large part formed the basis of my book.

FOREWORD

The author had a strong memory in 1971 when purchasing a house with a FHA mortgage at the end of the process, he was given a warning note as being inside of the 10-mile zone of the Rocky Flats, Colorado nuclear processing facility; but he was provided with virtually no further information. What may have been possible consequences including the risk to resale value? Caution is urged when assessing the nuclear industry and the attendant environmental effects. The following pages are intended to be of educational value and documentation of the past and future.

This book is divided into four major parts: Introduction to the Manhattan Project; Uranium Mining and Milling; Uranium Enrichment Facilities; and selected Nuclear Weapons Facilities- six in the United States and one in Siberia, Russia. The period covered is some 80 years from around World War II to the present. This information is directed at the general reader who is interested in science and wishes to dig deeper into the world of nuclear manufacturing and radiochemistry. Chapter 1 provides the early history of the Manhattan Project. Chapter 2 gives a short treatise of radioactivity. We will look at the massive mining and milling operations of uranium ores principally in the Southwest U.S. in Chapters 3 and 4. Chapter 5 deals with the enrichment of uranium-235.

In Chapters 6 through 12 are described the previously secret operations at the very large Soviet Mayak nuclear plutonium manufacturing plant in the Urals- the answer to the U.S.'s Hanford, Washington plutonium facility. Next, in the 1940s to the present, the focus is on the four leading nuclear weapons facilities of Oak Ridge, Tennessee, Los Alamos, New Mexico, Hanford, Washington and Rocky Flats, Colorado. Additional nuclear plants are covered in lesser detail including Pantex, Texas and Savannah River, South

Carolina. Other military and commercial nuclear plants will not be covered. The adverse consequences to the environment of each of these places caused by long-term processing followed by their cleanup activities will be covered as much as the available information permits.

The author's view of the nuclear weapons industry and the nuclear power industry is in part described in the following. It is recognized there are many and diverse viewpoints and some of the subject material is controversial. The Epilogue provides thoughts on waste cleanup strategies and final comments.

- The pale of confidentiality imposed over nuclear facilities and their workers for many decades meant that particular radiological events and worker health issues were not made known to the public and to the occupational workers.
- When certain confidentiality was lifted by Secretary of Energy Hazel O'Leary in 1993-1994, it nevertheless continued to persist in many cases. Confidentiality is the direct opposite of accountability which was foreign to the nuclear industry particularly in the earlier decades.
- It is known that records were not being kept in many operational areas or were lost, especially at the earlier nuclear facilities.
- The author believes nuclear annihilation is an ever-increasing concern and a major and real threat to the world. It is thought that Iran will likely achieve its goal of developing an atomic weapon by the end of 2023. Iran is thought to be presently nearing weapon grade uranium-235 levels.
- The media on the Manhattan Project may have been encouraged to minimize the importance of certain events.
- The development of nuclear power and nuclear weapons over the past 80 years has been controversial, and perhaps should have been subject to more checks and balances including ensuing environment cleanup.

- The U.S. is proposing a future system of measuring environmental damage as part of economic metrics. Other nations have incorporated "nature" into their core accounting, but the U.S. has minimized this approach. Systematic risk should be more fully assessed. The author believes equity is needed in this area, including that within the nuclear industry, but this issue has been contentious.

Table of Contents

PART ONE

INTRODUCTION TO THE MANHATTAN PROJECT

1

THE EARLY MANHATTAN PROJECT AND TODAY

Early Discoveries and Start of the Manhattan Project

In early 1932 at the University of California at Los Angeles, a chemistry student- Glenn Seaborg heard about the discovery of the neutron by the British physicist James Chadwick. In the fall of 1934, Seaborg entered graduate school at Berkeley, California where both Ernest Lawrence- the inventor of the cyclotron and J. Robert Oppenheimer were present. A cyclotron is a machine that accelerates charged particles or ions to high energies for investigating nuclear structure. At that time, the heaviest elements in nature like uranium and radium were known to be unstable and emitted particles like electrons, alpha particles and gamma rays. Seaborg using the cyclotron, was bombarding different elements with fast-moving particles and producing new radioisotopes like cobalt-60 used to later treat cancer and sterilize medical instruments, technetium-99 for imaging the liver, lungs and brain, and io-dine-131 to diagnose and treat thyroid problems. Then he began looking into the behavior of uranium and other heavy elements when bombarded by subatomic particles.[1]

In January 1939, the news was out that German scientists had bombarded uranium atoms with neutrons, splitting the atoms nearly in half, and releasing immense amounts of energy, a process that later was called fission which was a chain reaction with neutrons producing more neutrons and more fission. The question was asked by Seaborg and others as why chain reactions did not occur in uranium ore, which was answered by experiments at Columbia University in early 1939 where they found that only U-235, a tiny fraction in the uranium ore, would itself fission and produce a chain reaction.[1]

Physicists discovered that U-235 was splitting into barium and krypton and into all sorts of other atoms including xenon, strontium, iodine, yttrium, cesium and rubidium- more than 20 different elements. These fission products tend to be highly radioactive. The physicists realized if enough U-235 could be obtained from natural uranium ore, nuclear fission could be achieved and a controlled reaction could produce virtually unlimited energy, whereas uncontrolled reaction could produce a bomb of unprecedented power. A way was needed for converting U-238 to U-235 and to slow down the neutrons so that they would not be absorbed by the U-238. Carbon or pure graphite was selected for the purpose of slowing down the reaction so neutrons could be absorbed by the U-235. In early 1940, Columbia University received grant money to purchase a large quantity of uranium ore and extremely pure graph-ite. The researchers created a reactor which at that time was called a pile. At this point, the race was on between Germany and the U.S. to see which country would develop atomic bombs first. In June 1942, the Manhattan Engineer District was created and became known as the Manhattan Project. The scientists were asked to demonstrate the feasibility of a chain reaction and design a pilot plant to produce plutonium, which was accomplished at the University of Chicago in December 1942. This led to the highly desirable weapons grade isotope- plutonium-239.[1]

The Manhattan Project was a research and development project that produced the first nuclear weapons during WW2. The Project was under the direction of the U.S. Army Corps of Engineers and Major General Leslie Groves. The Project started slowly but grew to employ more than 130,000 persons and cost $2 billion spread across thousands of square miles of federal reservations and a network of 30 sites in the U.S., Great Britain, and Canada. The first bomb dropped on Japan was a uranium bomb using U-235. It is noted that uranium is relatively common on earth, and uranium ores generally contain from 0.3 to 1.0% uranium. U-238 makes up 99.3%, and U-235 only 0.7% of natural uranium. Uranium mining and milling was developed to produce yellowcake, as a large-scale industry mainly in the Western U.S., which had significant environmental consequences, as shown later in this report. Early-on, the Belgium Congo had the world's richest supply of uranium ore, which the Germans were trying to exploit. In the U.S., there was the danger that uranium ore shipments from Canada might be halted, and Colorado supplies were being considered.[2,3]

Separation of uranium-235 from uranium-238 was difficult, and was achieved by electromagnetic, gaseous and thermal means. In parallel with the work on uranium was the effort to produce plutonium which was used in the second atomic bomb dropped on Japan in August 1945. Major reactors were constructed at the Oak Ridge, Tennessee and Hanford, Washington nuclear facilities for irradiating uranium and producing very small amounts of plutonium-239 from very large amounts of uranium. The Manhattan Project was a fast-track program involving utmost secrecy in an effort to beat the Soviet Federation and Germany in atomic bomb development.[2,3]

It was said that uranium could provide bombs with a greater destructiveness than anything then known, and the nuclear weapons race was on. In September 1942, land was acquired for the start of the Oak Ridge, Tennessee facility. The Los Alamos, New Mexico Laboratory under the Manhattan Project was established in 1942 as the heart of the program responsible for design and assembly of nuclear weapons under the direction of J. Robert Oppenheimer. The Hanford, Washington plutonium production site was started in December 1942 to early 1943. The plutonium reactor called X-10 was started at Oak Ridge in March 1943 and the S-separation plant came on-line in July 1944. 59 kg. of uranium enriched to 85% was delivered by Oak Ridge to Los Alamos. Uranium slugs arrived at Hanford in March 1944, and irradiation of the slugs commenced in August 1944 with Reactor B. That reactor went critical in September 1944. Subsequent reactors at Hanford were brought online in 1944 and 1945. Work at Hanford proved dangerous, and by the end of the war, half of the experienced chemists and metallurgists had to be removed from work with plutonium because unacceptably high levels of plutonium appeared in their urine.[1,2]

Besides the Los Alamos Laboratory, other laboratories included the Lawrence Berkeley Laboratory, the Oak Ridge Laboratory, Argonne Laboratory, Ames Laboratory, the Brookhaven Laboratory at Camp Upton, Long Island, and the Sandia Laboratory at Albuquerque, New Mexico. The Lake Ontario Ordnance Works (LOOW) near Niagara Falls, New York became the principal repository for wastes from the Manhattan Project for the eastern U.S. All radioactive materials stored at this site including thorium, uranium and the world's largest concentration of radium-226 were placed into an "Interim Waste Containment Structure" in 1991 which was a burial site.[1,2]

In 1944, there was a movement among some scientists, including the renowned Neils Bohr, who advocated that atomic power should be used for the benefit of all mankind, and not become a menace to society. The Chicago scientists who accomplished the first uranium chain reaction, Oppenheimer-head of the Los Alamos, New Mexico Lab, and other scientists argued that Russia should be informed of the Manhattan Project to prevent an arms race, and the public should also be informed. Roosevelt, Churchill, and Groves strongly believed otherwise. In June 1944, Churchill, who had a strong influence on Roosevelt, signed an agreement for controlling supplies of uranium and thorium ores over the world, especially in the Congo. Churchill was fearful Britain might go bankrupt after the War, and by the end of 1944,

Roosevelt was said to have the dream of amassing overwhelming military power. At this point, the U.S. leaders and military were also thinking of developing the hydrogen bomb, more than 1,000 times stronger than the Hiroshima bomb.[2,3]

The first atomic weapons test took place at Trinity, near Alamogordo, New Mexico on July 16, 1945. The bomb was detonated atop a tower in the desert of New Mexico, about 150 miles south of Los Alamos. Testing at the Nevada test site began in 1951, and a total of 928 explosions would be conducted there before the site was closed in 1992. The mushroom clouds of nuclear explosions were seen in towns as far as a hundred miles distant. In Las Vegas, 65 miles away, the citizens dreaded the earthquake-like explosion shocks, whereas tourists enjoyed spectacular views of radiation clouds. Some of the Nevada fallout reached as far as New York State. Government representatives said the harmful effects of radiation were exaggerated.[4]

Hiroshima and Nagasaki, Japan

Kyoto in Japan was tentatively selected as the first city to be hit by the atomic bomb, but Secretary of War Stimson rejected this choice because of its cultural significance. Kyoto with a population of about 1 million was more than 3 times the size of Hiroshima and 4 times the size of Nagasaki, and if it had been successfully bombed, the death toll could have easily been ½ million. Groves strenuously objected, but Stimson prevailed with Truman's support. Hiroshima was selected as the first target.

Nagasaki was a late selection and did not meet the criteria of the Target Committee. It was a center of Western learning and had a large Catholic population. As Groves reflected after the war, Truman did not have to issue an order to use atomic bombs on Japan. The endgame had been determined long before when Roosevelt backed the Manhattan Project. Groves said Truman "was like a boy on a toboggan." Truman and his advisors had accepted the idea of possibly killing hundreds of thousands of Japanese civilians. With the

Trinity test, most observers remembered the intense bright light from the bomb, many times brighter than the sun. Groves early on had plans to drop two such bombs on Japan.[1]

In the final year of World War II, the Allies prepared for a costly invasion of the Japanese mainland. This undertaking was preceded by a firebombing invasion of the Japanese mainland that devastated 64 Japanese cities. The war in Europe concluded when Germany surrendered on May 8, 1945, and the Allies turned their full attention to the Pacific War. By July 1945, the Allies' Manhattan Project had produced two types of atomic bombs: "Little Boy," an enriched uranium gun-type fission weapon, and "Fat Man," a plutonium implosion-type nuclear weapon.

The Hiroshima atomic bomb named Little Boy with a force of 29 kilotons was dropped on August 6, 1945 (Figure 1.1); and only three days later on August 9, 1945, the plutonium bomb called the Fat Man with a force of 16 kilotons was dropped on Nagasaki.

Figure 1.1 Atomic Bomb Destruction of Hiroshima, Japan
Credit: National Archives

At Nagasaki, the bomb's shock wave destroyed virtually every building within a mile and a half of the hypocenter. On August 15, 1945, the emperor of Japan ended the war. The short-term effects of radiation on Nagasaki residents were devastating. The data collected on radiation sickness by the Nagasaki medical people were invaluable in understanding the effects of the atomic bomb on humans, but this data did not become publicly available for years. When the U.S. Army took control of Japan in September 1945, it imposed strict censorship on the dissemination of such information. U.S. scientists had largely failed to anticipate and evaluate the effects of radiation sickness, and the military did not want that information to be known. Groves dismissed such stories as propaganda, and said the atomic bomb was not an inhumane weapon. Censorship applied to everyone including American reporters. In Nagasaki, funeral pyres burned for months as bulldozers scraped the valley clean. The psychological damage to the people was as grave as the physical damage.[1,2,3]

By today's standards, the atomic bomb dropped on Nagasaki was small. The most powerful weapon in the current U.S. arsenal can release more than 50 times the energy of the bombs dropped on Japan. If a single hydrogen bomb were detonated above any city in the world, it would destroy much of that city. Houses within 5 miles of the hypocenter would be smashed, and persons 6 ½ miles away would suffer life-threatening burns. Today even with nuclear weapons reductions, multiple warheads target most American and Russian cities. After a large-scale exchange of nuclear weapons between two superpowers, few people would emerge from these cities alive. Then the cities would burn. Unprecedented climate changes would threaten the global food supply and temperatures would be reduced so much, that humans surviving the initial bombing would starve. This situation would dwarf the effects of climate change visualized by most people in the world today. The immediate threat is in nuclear weapons sitting in missile silos, bombers, and submarines around the world.[1]

The stage was set for an international arms race. In August 1946, the Atomic Energy Commission (AEC) took over the program from the U.S. Army Corps of Engineers although the Manhattan District was not formally abolished until August 1947. There was general agreement that after the war, the atomic energy program of the U.S. should remain intact, that a sizeable stockpile of material for military use should be maintained and the door for industrial development should be opened. There was a strong contention if Washington rather than the military had maintained better control of the scheduling of their bomb raids on Japan, the destruction of Nagasaki could have been prevented. Many said that the first bomb may not have been necessary, but the second one certainly was not. During this critical period, the Japanese were seeking concessions that the emperor and the dynasty be spared. The nuclear destruction of Hiroshima and Nagasaki showed the most important consequence- that the survival of humanity could no longer be assumed.[1,2,3]

One scholar said whether the atomic bombs dropped on Japan could have been avoided was the most controversial issue in all American history. The consensus view of most historians is that an invasion of Japan probably would not have been necessary to end the war, especially if the atomic bombs were not ready to drop on Japan in August 1945. Most Americans believed differently, and they were influenced by a strong postwar public relations campaign carried out by officials involved in the decision. Whether dropping atomic bombs on Hiroshima and Nagasaki was necessary will remain one of the great "what ifs" in human history. The person who discovered plutonium- Glenn Seaborg had written to Ernest Lawrence not to use the weapon on Japan without warning. The U.S. itself has never renounced the first use of nuclear weapons, even if their use risked the end of human civilization. As reports came back from Hiroshima and Nagasaki, Oppenheimer and many of his colleagues at Los Alamos became increasingly depressed. The Nagasaki bombing just 3 days after the first atomic bomb on Hiroshima and before the Japanese leaders had the chance to fully assess the damage and their options, was troubling. The physicist who called out the countdown for the Trinity test said when the second bomb was released, "we felt it was a great tragedy." Leo Szilard in Chicago, the first man who conceived how an atomic bomb might work, wrote a friend saying its use against Japan was "one of the greatest blunders of history."[1]

Testing at the Pacific Bikini Atoll and Nevada Test Site

On April 25, 1953, an atomic bomb was detonated at the Nevada test site with a power much greater than predicted- almost three times the power of the Hiroshima bomb, at 43 kilotons, and knocking down VIPs and former congressmen. The story was to trust the government spokesmen. The medical people did not seem much concerned. The yield of bombs detonated at the Nevada test site was limited to 1 megaton, but President Truman was looking for a new superbomb- a fusion hydrogen bomb, and thus the testing had to be moved to the Pacific. Many scientists were opposed, including Robert Oppenheimer, the head of Los Alamos. A prototype hydrogen bomb had already been detonated on Bikini Atoll in the Marshall Islands in November 1952.[4]

The prototype hydrogen bomb had a yield of 10.4 megatons, more destructive power than the planet had ever experienced. Elugelab Island, which was ground zero, totally disappeared from the earth. The radioactive dust of vaporized coral descended on ships dozens of miles away. A censored film documenting the execution of the test was released to the public on April 1, 1954. The government said the bomb cloud reached an altitude of 125,000 feet, and the major portion of atomic debris was carried well into the atmosphere. AEC experts believed the stratosphere could be a dumping ground for nuclear fallout. The narrator of the film mentioned there was potential danger to some 20,000 people living in the area. In October 1953, the agreement was made that 7 tests would be carried out on Eniwetok Island in the Pacific Proving Grounds. The first of these tests would be called Castle Bravo with an estimated yield of 6 megatons, and it took place on March 1, 1954. Everything was under complete secrecy. General Clarkson had 12,945 men,

24 ships including the USS Estes, 129 barges and boats, and 76 aircraft. The objective was to drop a plane-delivered bomb called the "shrimp."[4]

The winds at the Bikini Atoll test site were becoming somewhat less predictable, and two days before the test, no ships were found that did not belong in the area. The scan area was roughly 200 miles by 800 miles. The shock wave from the March 1 explosion rocked ships in the 50-mile zone from side to side. A report signed by General Clarkson said the test was highly successful. But inside the observation bunker on Enyu Island, the radiation unexpectedly continued to rise both outside and inside the bunker, and the same was being recorded on ships within the 50-mile zone. On the destroyer USS Phillip, radiation levels were up to 20,000 milliroentgens/hour; the USS Bairoko registered readings as high as 25,000 milliroentgens/hour; and all ships were ordered to get out of the area immediately. The lead person in the observation bunker was rescued and underwent scrub down procedures. It was estimated that fallout outside the blockhouse was as high as several hundred roentgens. There was extensive contamination throughout the region, and the magnitude of the explosion was estimated at 15 megatons- a thousand times larger than Hiroshima.

Evacuation of the entire population of Rongelap Atoll and associated islands- more than 100 miles from the bomb site, was ordered in the late pm of March 2 but did not start until 48 hours after the test. Unfortunately, until the evacuation, the islanders received no notification of the radioactive fallout, and the physical damage had already been done. On the pm of March 2, two U.S. officers had arrived and taken radiation measurements which showed the villagers' houses were exceedingly high at 1.4 roentgens/hour. Maximum dosage allowed for U.S. military personnel at that time was 3.9 roentgens over 3 months. On March 3, radioactivity at Utirik Atoll some 300 miles from Bikini was at 160 milliroentgens/hour. It was assumed these people if not evacuated could accumulate up to 58 roentgens of radiation- 15 times the dosage of 3.9 roentgens allowed. Over the next few weeks, burns, nausea, skin lesions et. al., were noted among the evacuees. Those caught in the fallout at Rongelap absorbed up to 130 roentgens, or more than 33 times the acceptable dosage levels. With the evacuations, General Clarkson was eager to proceed with the next six tests. On March 11, the AEC approved the evacuation of personnel from the islands due to nuclear fallout, but it assured the public there had been no harmful effects. The explosion of March 1 had been "a routine atomic test." This became an exercise in damage control, "friendly" journalists were there to help, and much of the information became confidential.[4]

The U.S. coverup was blown away by revelations made on March 16 regarding the 82-foot Japanese tuna ship- the Lucky Dragon No. 5, which returned home to Honshu. The crew of 23 had been away a month, fishing in the waters off the Marshall Islands. The Lucky Dragon was more than 70 miles from ground zero and some 25 miles outside the danger zone defined by the U.S. Navy but experienced very heavy nuclear fallout. It was missed by U.S. reconnaissance. The exposure of the Japanese fisherman was like that of those on Rongelap, and they knew nothing of what happened until they reached Japan. On arriving, levels of gamma radiation were up to 45 milliroentgens/hour but could have been much higher. It was estimated the fishermen sustained more than 20 yearly limits of radiation exposure. The U.S. occupational limit stands today around 4.4 roentgen. Japanese doctors reported these fishermen experienced symptoms like the survivors of Hiroshima and Nagasaki. Everyone and everything were quarantined.

The irradiated food that the fishermen had eaten on board was part of their catch which the citizens on the mainland realized was contaminated. But some of this catch had already been eaten by the citizens. However, contaminated fish had also come from other ships returning from the Pacific. By the end of the year, 75 tons of tuna were destroyed in a seafood-eating country. In return, a U.S. committee of the Atomic Energy made the assertion that the Lucky Dragon had been within the restricted zone and on a spy mission. The chairman of the AEC- Lewis Strauss said "the wind had failed to follow the predictions," and he downplayed the situation. U.S. refusal to release technical information meant the subject continued to be discussed for the rest of the decade. Los Alamos had proved inaccurate in predicting the yield of succeeding bombs with serious underestimates because they failed to understand the behavior of lithium-7 used in the bombs which transformed into tritium in the fusion reaction. General Clarkson complained that the radiological standards to protect personnel were still too restrictive, and he was granted the right to waive the limits to complete his operations.

General Clarkson of the U.S. Army and the AEC were said to be more than happy to put the Castle Bravo event behind them, and security and the confidentiality that was maintained for a long time further clouded the picture. In the 1950s, Castle Bravo became internationally a symbol of imperialism and colonialization. Nuclear testing was a "hot" topic. Albert Einstein and Bertrand Russell signed a letter expressing the suicidal nature of nuclear war. It was said a bomb could now be manufactured which would be 2,500 times as powerful as that which destroyed Hiroshima. The Rongelap people were exposed to sustained radiation, and almost every child who was exposed to Castle Bravo fallout ended up with a thyroid problem including tumors, abnormalities and even death. For decades, the U.S. offered considerable money for cleanup of irradiated waste and dealing with the medical, social, and economic issues faced by the Marshallese people. The publicity over Castle Bravo was more than a wakeup call to the world of the consequences of the hydrogen bomb.[4]

Post War, Cold War and U.S. Secretary of Energy O'Leary

In the early-1950s, anti-nuclear activism was depicted as being pro-communism, and with the Korean War and McCarthyism, there was concern with the U.S. exploding a nuclear weapon in Korea, which was being pushed hard by the military. General MacArthur asked for authorization to use atomic weapons, and in his journal said it would take many atomic bombs to end the war in Korea. Truman was in favor as were Johnson and Eisenhower later. Leaders around the world cautioned Truman. In March 1950, a Committee for World Peace in Stockholm, Sweden had previously demanded a ban on all nuclear weapons. Many churches in the U.S. rejected such a ban as did the American and British Federation of Atomic scientists because to do so would support atheistic communism. McCarthyism gained tremendous momentum and the creditability of the U.S. was seriously compromised by the prevailing justice system. It is said historians failed to discuss the courage it took to protest nuclear weapons and the Korean War.[5]

The story of the American atomic bomb was born in secrecy and continued that way for a long time. This nuclear secrecy continues today, and science has been put at risk because it has not been made free. Such secrecy has been controversial and contested and made the American public dangerously ignorant of the evolving national and world situation. To hide the truth, the stories being told at Oak Ridge and Hanford were misleading. Threats from the FBI were common for those tampering with Groves' censorship rules. It was said the AEC morphed into one of the most secretive bureaucracies in U.S. history. The AEC stayed low-key until three shocks came in late 1949 to early 1950: detection of the first Russian atomic bomb; revelation of Soviet penetration of the Manhattan Project; and the acrimonious H-bomb debate. From 1947 to 1949, the AEC had the FBI investigate over 140,000 individuals for top-level security clearance- approximately one out of 800 adult Americans.

America seemed to be shocked they no longer had a monopoly on the atomic bomb. Detonations at the Nevada test site could sometimes be seen as mushroom clouds at the casinos in Las Vegas. The Bravo test in the Marshall Islands in the Pacific exposed many Marshallese people to high levels of radiation, together with many American servicemen and occupants of a Japanese fishing vessel in the area. When radioactive tuna entered the Japanese market, that country panicked. The head of the AEC said… "at no time was the testing out of control." The fact that Bravo produced massive fallout contamination was known not just to the newspapers, but to the Soviets as well.[6]

Hazel O'Leary, U.S. Secretary of Energy (from 1993 to 1997) said in 1947 that 18 Americans previously diagnosed with terminal disease, had been injected with solutions of plutonium under minimal standards of informed consent to better understand human absorption of plutonium for the purpose of establishing plant safety standards. This lack of consent was controversial in 1947 when the AEC placed it under "restricted data", rather than having to deal with its adverse publicity. It was found that some 800 total human radiation experiments were conducted under inadequate standards of consent as of that time. Even after tomes of declassified reports were released to the public by the DOE, Congress in late 1988 passed a new law requiring DOE to review all declassified documents for possible "inadvertently released" restricted data. But in 1994, O'Leary said the DOE was now going to release information on long-term environmental damage during the years of nuclear testing and contamination at the Rocky Flats, Colorado plutonium fabrication plant.[5] It was then learned that from the 1940s into the 1970s, the U.S. government subjected more than 23,000 people to radiation experiments, many without their informed consent. There were also experiments on pregnant women. In the 1950s and 1960s, most workers at the Fernald nuclear weapons plant near Cincinnati Ohio were overexposed to uranium without their knowledge or consent. At nuclear weapons plants, workers were falsely assured that they were not being harmed.[7,8,9]

Confidentiality

The AEC was trying to classify anything produced by private companies relating to nuclear material studies as "restricted data", even if that company had wholly developed the procedures on its own, and then they had to get permits from the AEC to continue their work. This situation continued until the AEC was terminated in 1977. President Carter indefinitely suspended reprocessing of nuclear materials in part because of security problems in an ever-increasing amount of separated plutonium being in civilian hands. The bottom line, however, was- it is relatively easy for an unsophisticated chemist to make an atomic bomb with horrifying consequences. In the 1980s, the Natural Resources Defense Council (NRDC) tried to stop development of certain types of nuclear reactors and started to provide the basic facts about U.S. nuclear weapons. William Arkin of the NRDC reported the U.S. had placed tens of thousands of nuclear weapons in dozens of countries, some-times without knowledge of their high commands, and with no knowledge by their inhabitants. The State Department looked into criminal charges against Arkin, but he was exonerated. Several technical books then came out on the specifics of the manufacture, location and power of nuclear weapons.[6]

Alex Wallerstein has said… America is a place built upon values in conflict- a simmering mix of high-minded idealism and ugly, fearful power. There continues to be in the U.S. a pervasive, conspiratorial, bi-partisan suspicion of secrecy in thought, and there are deep questions about the legitimacy of government claims to control information in the name of national security. Secrecy continues and there is no sign that it is going away. Knowledge is power, especially if you are in the "inside," but also, that information is difficult to control. However, an outsider once privy to secret information, can completely change his/her perspectives.[6]

Continuing Nuclear Events

By the end of the 1950s, the Soviets, the U.S., and the U.K. exploded an estimated 307 nuclear weapons. In October 1962, Soviet nuclear missiles were under construction in Cuba, and Russia and the U.S. came very close to war. U.S. subs were trailing Russian subs that were trailing the ships carrying the atomic weapons to Cuba. Production of the MX missile- a heavyweight ICBM which could carry 10 nuclear warheads was stopped in December 1982, but that of the Trident and Trident II, the Pershing II missile, Cruise missiles and anti-satellite weapons continued.

In the second half of the 1980s, the U.S. and Russia signed the START treaty and the Intermediate Range Nuclear Forces treaty. the Soviet Union broke apart in 1991, and some 6,000 atomic bombs possessed by Russia were spread over many countries. The U.S. helped Russia and Kazakhstan to secure a better hold on these nuclear materials, but much of this went missing. A. Q. Khan, a Pakistani, helped his country achieve nuclear bomb status. Kahn continued to sell nuclear materials to various countries, many of which were

U.S. enemies including North Korea, Libya and Iran. Pakistani scientists were also selling this material to al Qaeda. Kahn was arrested, but released in 2009, and these secrets had been released to the world.

In October 2011, the final parts of a massive B53 bomb were dismantled in Texas. The bomb weighed 10,000 pounds and was intended to be more than 600 times more powerful than the Hiroshima bomb. The U.S. defense budget has steadily gone up over the decades to the present. Sooner or later, an atomic bomb or two may likely be deployed.[5,10]

Death by plutonium was vividly illustrated by the bombing of Nagasaki in WW2. Some 70,000 deaths were attributed to the plutonium explosion caused by the Fat Man bomb, and this number doubled after five years. This compared to another 200,000 deaths five years after the dropping of the Little Boy bomb on Hiroshima.[11]

By the winter of 1995 plans were underway for commemorating the 50th anniversary of the bombing of Hiroshima, Japan, and it was evident the wounds of that event still had not healed. Some 76% of Americans said the U. S. did not owe Japan any apologies, and 58% said there was nothing morally wrong with the bombings. Firebombing of many Japanese cities had preceded the atomic bombs dropped on Hiroshima and Nagasaki. The firebombing of Tokyo on March 9, 1945, destroyed 16 square miles of that city and killed between 34,000 and 100,000- some 10% of that city's population. Other cities that were incinerated included Nagoya, Kobe, Osaka, Yokohama and Kawasaki where as many as 400,000 Japanese were killed. Analysts have claimed that destruction of Hiroshima and Nagasaki prevented the loss of lives of many more persons if otherwise an all-out invasion of Japan had been carried out. Historians however discovered evidence that Japan was prepared to surrender before August 1945.[11]

Oversight Committee Reports

The Office of Technology Assessment (OTA) of the U.S. Congress released its evaluation of the nuclear industry in 1991. It was found that serious environmental problems were predicted more than 30 years previously; but too little attention was paid to its findings. Serious questions were also raised as to the potential human health threats from such contamination. The report said the 45 years of nuclear weapons production post WW2 caused the release of vast quantities of radionuclides and hazardous chemicals into the environment. And there was evidence that air, groundwater, surface water, sediments and soil, as well as vegetation and wildlife, had been contaminated at most- if not all- of the Department of Energy (DOE) nuclear weapons sites. At every facility, the groundwater was contaminated with radionuclides or hazardous chemicals. Many sites in non-arid locations have surface water contamination. Millions of cubic meters of these same wastes have been buried throughout the nuclear network, and records of burial site locations and contents are in most cases inadequate or may not exist. Total contaminated soils and sediments of all types are estimated to be in the billons of cubic meters. Also, very little quantitative characterization of each site had been done as of that time.[12]

The OTA report described in the early 1990s large quantities of radioactive and hazardous wastes in storage at all the sites, often under marginal conditions. There is an increasing need to store these wastes safely and for long periods of time until disposal alternatives are available. It seems this situation continues up to the present day- with burial being the most chosen solution.[12]

The Subcommittee on Oversight and Investigations of the Energy and Commerce Committee on May 2019 did not understand why with billions of dollars spent by DOE in 2018 for cleanup, the overall environmental liability for all nuclear sites had dramatically increased. The Subcommittee said DOE did not have an overall comprehensive plan for cleanup, and significant cultural change was necessary. The GAO further said it is difficult to tell whether the U.S. is getting value for our dollar, and emphasized contracting is not the same as project management. Yucca Mountain was mentioned as a wishful final repository for nuclear wastes, but does not seem in the picture for the foreseeable future. Vitrification (a process of rapidly cooling a liquid into an amorphous glass-like substance without crystals) had been used at the Savannah River nuclear facility to treat 35 million gallons of high-level wastes but the material remained sitting in South Carolina because Yucca Mountain was never approved for final disposal. It is noted that Savannah River is an ongoing nuclear weapons processing site.[11,13]

As reported in a March 2021 article on the Paradoxes of Deterrence, leading voices in the Catholic Church not least of which was Pope Francis, condemned the very possession of nuclear weapons in 2017, and the Pope reiterated this statement in his visits to Hiroshima and Nagasaki in 2019. Bishop Robert McElroy of San Diego claimed that the possession of nuclear weapons "is now condemned, regardless of the intention." And the former editor of America Magazine- Drew Christiansen, argued "we should cease to imagine nuclear weapons as tools for us to manage, but rather as a curse we must banish." One can denounce the use of nuclear weapons particularly those involving harm to civilians, imperiling peace in unstable regions, and terrorists acquiring these weapons. But until a peace can be reached based on trust and mutual disarmament, deterrence may be the only choice.[14]

2

RADIOCHEMISTRY

We will now get into a little bit of chemistry particularly as regards uranium and plutonium. Since the early 1900s, scientists knew that atoms contained positively charged particles called protons which are clustered together in a central nucleus, and this nucleus was surrounded by much lighter particles called electrons having a negative charge. The number of protons determines what kind of element the atom is. At that time, hydrogen was at the top of the list with one proton, and uranium was at the bottom of the list with 92 protons. People however wondered what kept the nucleus together since the protons would just repel each other. Later, the neutron was also discovered inside the nucleus. Neutrons having a neutral charge act somewhat like a sort of nuclear glue, sticking to the protons and keeping them from blowing apart. Also, atoms with the same number of protons can have different numbers of neutrons, which accounts for their different weights of what is known as isotopes of the same element.

In 1934, it was known that the heaviest elements in nature like radium and uranium were unstable. Their nuclei emitted particles such as electrons; alpha particles consisting of 2 neutrons and 2 neutrons; and packages of light known as gamma rays. They change from one element to another until they find a stable configuration of protons and neutrons. For example, the most common isotope of uranium, i.e., uranium-238, has 146 neutrons compared to 92 protons or a weight of 238. However, even an overabundance of neutrons is not enough to render heavy elements completely stable, and eventually they decay to lighter elements.[1] Figure 2.1 shows the decay transformation of the uranium- radium family, and Figure 2.2 gives relative occupational limits and half-lives for various members of the uranium-radium family.[15]

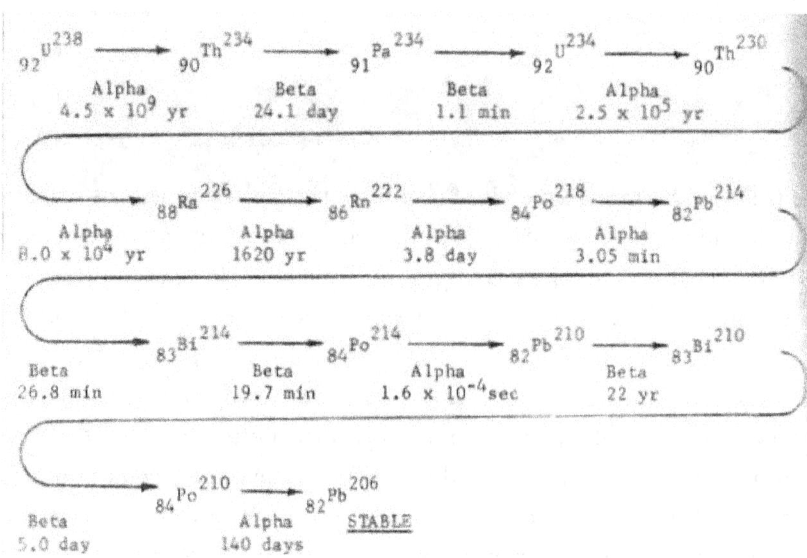

Figure 2.1 Decay Transformation of Uranium–Radium Family
Source: USDHEW, Region 8

Isotope	MPC$_W$ pc/liter	Critical Organ	Half-life	Emission
Ra^{226}	3.3	Bone	1,620 yr	Alpha
Pb^{210}	33	Kidney	22 yr	Beta
Po^{210}	233	Spleen	140 day	Alpha
Th^{230}	667	Bone	8×10^4 yr	Alpha
Th^{234}	6,667	GI tract	24.1 days	Beta
U^{234}	10,000	GI tract	2.5×10^5 yr	Alpha
U^{238}	13,300	GI tract	4.5×10^9 yr	Alpha
Bi^{210}	13,300	GI tract	5 days	Beta
Pa^{234}	2/	-	1.1 min	Beta
Po^{218}	2/	-	3.05 min	Alpha
Po^{214}	2/	-	1.6×10^{-4} sec	Alpha
Bi^{214}	2/	-	19.7 min	Beta
Pb^{214}	2/	-	26.8 min	Beta
Rn^{222}	(gas)	Lung	3.8 days	Alpha

1/ MPC$_W$ value is the maximum permissible concentration in water for average member of the general population (1/30th HB69 value for continuous occupational exposure).

2/ No value given for these short-lived materials.

Figure 2.2 Occupational Limitations and Half-Lives, Uranium-Radium Family
Source: USDHEW, Region 8

In 1934, the husband-and-wife team of Joliet-Curie discovered that *ALL* elements, have isotopes that are radioactively unstable. Bombarding an element with sub-atomic particles can create nuclei with unstable combinations of neutrons and protons that are artificial and do not exist in nature. These different collections of protons and neutrons stay around for a while, and they eventually decay further. Thus, radioisotopes with new, unusual and useful properties are created.[1]

Attention was then turned to bombarding uranium atoms with neutrons to produce a chain reaction with the release of immense amounts of energy. It was learned uranium consists almost entirely of two isotopes- one with 92 protons and 146 neutrons called uranium-238, and one with 92 protons and 143 neutrons called uranium-235, but it was only the latter that would fission, which however made up only 0.7% of the uranium ore. Raw uranium ore would not fission. Uranium-235 would split into 20 different elements altogether, most of which were highly radioactive.[1]

Knowing that uranium-235 would fission and create a chain reaction, the scientists then moved on to plutonium giving them a second way to create an atomic reaction. In a reactor, uranium-238 captures a neutron from the fission of uranium-235 atoms creating uranium-239 which has 92 protons and 147 neutrons which however is unstable. This atom then converts one of its neutrons into a proton creating an element with 93 protons and 146 neutrons called neptunium-239. But the latter is also unstable converting another one of its neutrons into a proton with accompanying gamma and electrons yielding plutonium-239. The new artificial element had 94 protons and 145 neutrons and proved to be very interesting. This new element generated more neutrons than uranium, meaning it could produce even more powerful nuclear weapons. Separation methods were developed to refine and produce plutonium at various nuclear facilities especially Hanford, Washington, and this is the direction the U.S. eventually took in its nuclear weapons program.[1]

Measurement of Radioactivity

The overall measurement of radioactivity is complex. We will attempt to describe it in very simplified terms for our needs. Radiation is energy/unit mass. Source radiation is different than that absorbed by the human body, and lastly there is radiological biological damage. For source radiation, in the past, the standard unit was the "curie" which was a very large amount of radioactivity. The International standard for source activity is the "becquerel" where 3.7×10^{10} becquerels = 1 curie. 1 becquerel = 1 decay/second given off by a source material.

It is ionizing radiation that is the most dangerous, which interacts with cells and DNA. All nuclear energy is ionizing and can change DNA. The duration over which the radiation is received is important. A heavy dose received over a short time can be more fatal than the same dose received over a longer time.

Absorbed radiation represents a potential effect on the human body. It is measured in "Grays (Gy)", or in the U.S., in "rads". A rad is the absorbed dose of ionizing radiation equal to 100 erg of energy/gram of material irradiated. 1 Gray = 100 rads.

Biological Damage or Effective dose is measured in sieverts and rems where 1 sievert = 100 rems. Geiger-Mueller counters and dosimeters measure in millisieverts, rems and other, and relate the number of disintegrations/second.

The old occupational worker standard was 500 millisieverts/year compared to the public being limited to 100 millisieverts/year (5 times less). The new occupational worker standard became 50 millisieverts/year compared to the public allowed to receive 1 millisieverts/year, which is close to natural background. Internationally, 20 millisieverts/year was recommended for workers, but AEC/DOE objected, so the level of protection stayed at 50 millisieverts/year. Today, the 5-year allowable biological damage dose for nuclear workers in the West is 100 millisieverts or 0.1 sieverts.

For plutonium, the earliest total body burden was set for workers at 5 micrograms, but in 1945, this was lowered to 1 microgram plutonium, and in 1949 the limit was further lowered to 0.5 micrograms. In 1977, the maximum occupational plutonium dose allowed for workers was 5 rems/year (50 millisieverts/year). 16 grams of plutonium have the radioactivity of 1 curie. Plutonium is a dangerous radionuclide to handle.[4,15,16,17,18]

PART TWO

URANIUM MINING AND MILLING

3

URANIUM MINING

Uranium Mines in the Western United States

Uranium mining is the essential front-end of the entire nuclear weapons and nuclear power industry. One observer has said uranium is the true dirty underbelly of the whole industry, and hundreds of thousands of people have been adversely affected by such mining. In this report the focus will be mainly on the uranium mines and mills in the Southwestern United States. The advent of nuclear weapons and nuclear power in the U.S. resulted in a full-blown exploration and a mining boom immediately after WW2, with uranium the most important commodity. U.S. production peaked from about 1948 to the early-1980s when uranium prices were at their maximum with very high subsidies from the Manhattan Project. Today, the U.S. obtains less than 10% of its uranium ores and concentrates from domestic sources, the remainder coming from several different countries, the main producers being Australia, Russia, Kazakhstan, Namibia, Uzbekistan and Canada. In the 1980s, if not before, when demand and uranium prices dropped, most uranium mines were shut down together with their accompanying mills and then abandoned by their owners.[20,21,22,23]

There were at least 15,000 mine locations with uranium occurrence in 14 Western states. Most locations were in Colorado, Utah, New Mexico, Arizona and Wyoming, with about 75% on federal and tribal lands. The large majority of the early mines were conventional mines- open pit, strip mining and underground mines. These mines generated large amounts of bulk waste material including bore hole drill cuttings, excavated topsoil, barren overburden rock, weakly uranium-enriched waste rock, and subgrade ores. At some abandoned mine sites, ores enriched with uranium were left onsite. Today much of the uranium mining is done by in-situ or in-place leaching where the solution-bearing uranium is brought to the surface for transport and further processing. Uranium mining in the West caused considerable environmental degradation, and the cleanup of abandoned uranium mines (together with old, abandoned gold and silver mines) in the West represents an enormous problem.[20,21,22,23]

Much of the ores in the Western U.S. were low-grade containing only about 0.1% to 0.3% uranium. Rich uranium deposits found today worldwide may be as high as 2% to 25% in uranium. Of this uranium in typical western U.S. ores, only about 0.7% is uranium-235 which is the prized commodity. Therefore, enormous amounts of ore had to be mined to get the desired uranium. A major problem with these mines was keeping the underground operations from flooding, and pumps were in widespread use. Piles of so-called waste rock can contain elevated concentrations of radioisotopes. These piles and the mines can release radon gas and seepage water containing radioactive and toxic materials. The more modern in-situ mining also produces waste sludges and wastewater when recovering the leaching solution.[24,25]

Location of Uranium Mines by the USEPA and USGS, 2006

The EPA was tasked to address hazards posed by "technologically enhanced naturally occurring radioactive materials" or TENORM (see below).

Accordingly, the EPA investigated the environmental hazards of wastes from abandoned uranium mines operating in the United States. Between the 1940s and 1990s, there were thousands of such mines operating in the U.S., mostly in the Western U.S., many leaving a legacy of radiological and chemical hazards. The EPA compiled mine location information from federal, state, and tribal partners especially the Navajo Nation, to develop a unified database that could be used with GIS software. The number of mine locations associated with uranium as of 2006 or before, was around 15,000. Uranium mines have the potential to become health hazards if not properly closed. Three uranium mines were on the Superfund list while others were in the EPA CERCLIS (Superfund) hazardous waste database. The database did not reflect the current reclamation status of the approximately 15,000 uranium locations.[20]

Some definitions are needed in this area. Technologically Enhanced Naturally Occurring Radioactive Materials (TENORM) are naturally occurring radioactive materials that have been concentrated or exposed to the environment because of human activities. "Technologically enhanced" means the radiological, physical, and chemical properties of the radioactive material have been altered in a way that increases the potential for human and environmental exposure. Naturally Occurring Radioactive Materials (NORM) are materials which may contain any of the radioactive elements as they occur in nature, such as radium, uranium, thorium, potassium, and their radioactive decay products, that are undisturbed because of human activities and are considered "natural background radiation."

"Source Materials" are defined as 1) uranium or thorium, or any combination thereof, in any physical or chemical form; and 2) ores which contain by weight one-twentieth of one percent (0.05%) or more of uranium, thorium or any combination thereof. The Nuclear Regulatory Commission (NRC) also defined "Byproduct Materials" or wastes of these operations as tailings or wastes produced by the extraction or concentration of uranium or thorium from any ore processed primarily for its source material content, including discrete surface wastes resulting from uranium solution extraction processes.[20] The relation between "source materials" and "alternate fuel materials" being received at the White Mesa mill in Utah- the last remaining uranium/vanadium processing mill in the U.S. today, is not clear.

Up to the first half of 2005, the U.S. industry produced over 358,000 metric tons of uranium (U_3O_8). The EPA Uranium Location Database Compilation Report of August 2006 contains three illustrations of importance. Figure 3.1 shows Western uranium locations from the EPA Uranium Location Database as of 2006. Figure 3.2 shows Western uranium locations from the USGS MRDS Database. Figure 3.3 shows the density and cluster number of Western uranium mines spread across the Western states. An unfortunate legacy of uranium exploration, mining and ore processing was the numerous land workings left behind. Thousands of miners, prospectors, and large mining companies searched the United States for the valuable uranium deposits, like the California gold rush more than 100 years earlier. In many cases, they left behind waste with elevated radioactivity from uranium and its radioactive decay products, exposing people and the environment to its hazards. Most uranium mining in the United States was found in the expansive Colorado Plateau region including the Four Corners area and in Wyoming. But uranium mining also occurred in other areas of the United States.[20,23]

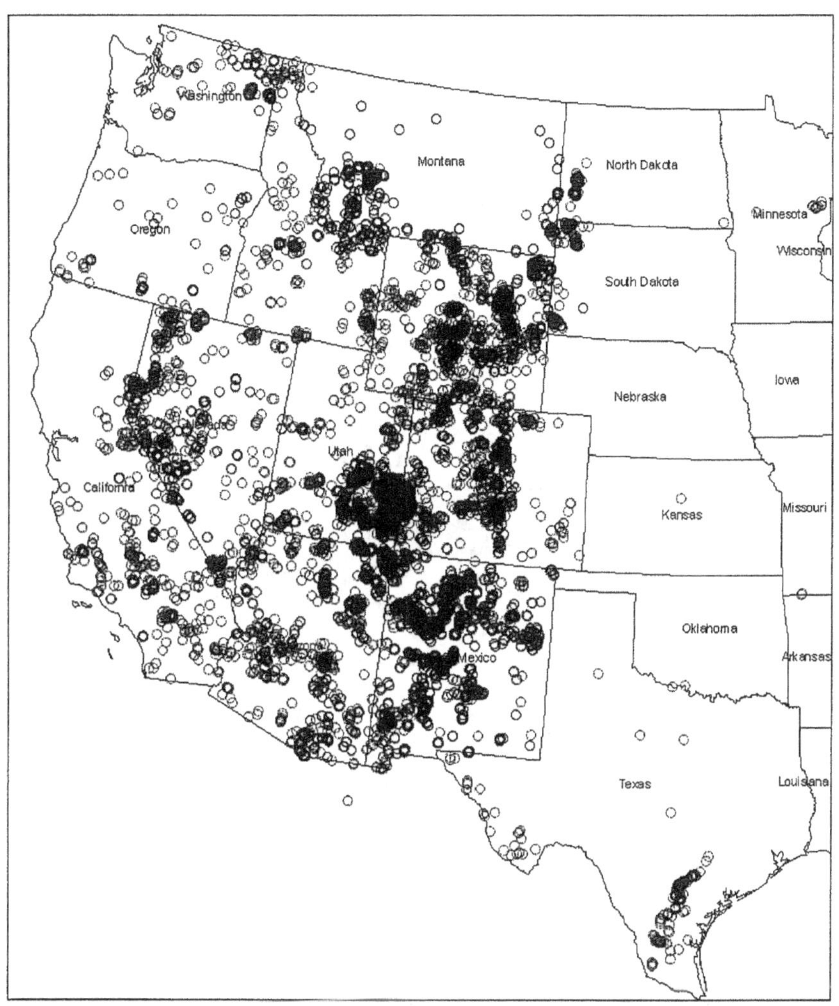

Figure 3.1 Western U.S. Uranium Mine Locations
Source: USEPA

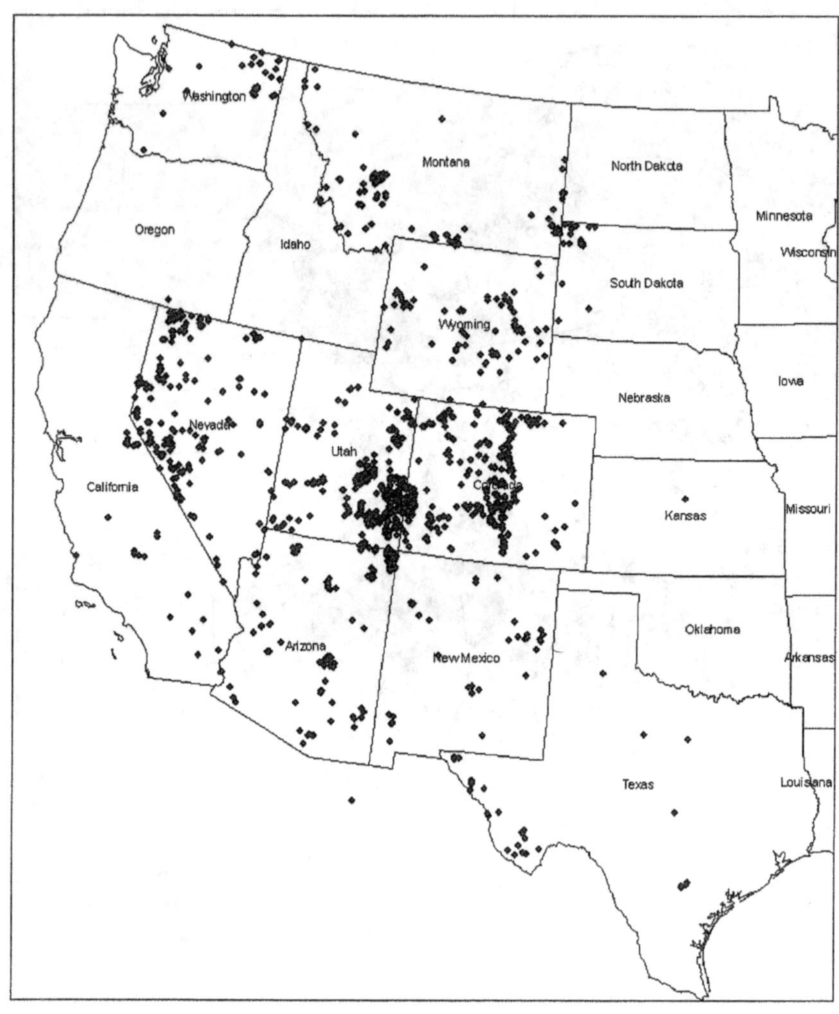

Figure 3.2 Western U.S. Uranium Mine Locations
Source: USGS

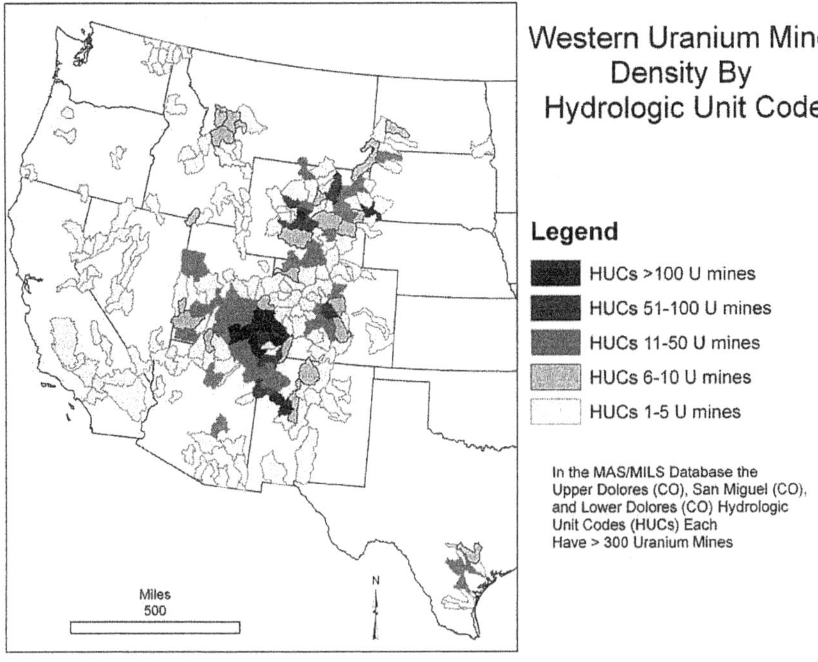

Figure 3.3 Density of Western Uranium Mines

Source: USEPA

Trump Administration Proposes $1.5 Billion for Mining in the West, February 2020

President Trump's $1.5 billion proposal to Congress to prop up the U.S. nuclear fuel industry captured interest for at least one company- the Toronto, Canada-based Energy Fuels Inc, to take steps to boost operations at dormant uranium mines in the West, including opening a mine 15 miles outside Grand Canyon National Park. The $1.5 billion project would extend over 10 years, build up a U.S. uranium stockpile, and reverse over-reliance on foreign uranium that harmed vital U.S. energy security. In February 2020, Energy Fuels Inc. announced a stock sale and said it would invest nearly $17 million for uranium mining in Utah, Wyoming, Arizona, Texas and elsewhere.

Energy Fuels has been one of the major companies asking for U.S. taxpayer support for domestic uranium mining which helped sell the Trump administration on reducing the size of Bear Ears National Monument in Utah and open up more land for possible future mining and oil and gas development.[25]

Residents near some of the mines in Utah said they feared an increase in radioactivity threats. White Mesa Utah, some 20 miles south of Blanding, Utah is home to 290 persons who are members of the Ute Mountain Indian Reservation close to the vanadium/uranium processing mill owned and operated by the Energy Fuels company. The Ute community protested practices at the White Mesa mill. Energy Fuels Inc. said the Trump proposal would lead to the production of 2.5 million pounds of uranium per year compared to 174,000 pounds produced in 2019.[26]

Energy Fuels, Inc. Proposal to Shrink Bear Ears National Monument, December 2017

Energy Fuels Resources urged the Trump administration to limit the Bear Ears Monument in Utah to the smallest size needed to protect key objects and archaeological sites, to make it easier to access the radioactive uranium ore (from Energy Fuels' dormant Daneros mine), and help it operate its nearby uranium processing mill located at White Mesa, Utah.[27]

Uranium Mines and Mills on Native American Lands, Radioactive Colonialism, 1992

For decades during the Cold War, uranium mines on Navajo Nation lands produced uranium used in nuclear weapons. On the Navajo reservation which stretches over Utah, New Mexico and Arizona where 300,000 persons reside, water is in short supply. In 1974, it was claimed that all federally controlled uranium production was originating from contemporary Indian reservation lands. In 1975, there were 380 leases containing uranium extraction ores compared to only 4 on non-Indian lands.[28]

Shiprock, New Mexico

In 1952, with the advice of the Bureau of Indian Affairs (BIA), and with the promise of jobs and royalties, the Navajo Tribal Council made an agreement with the Kerr-McGee Corporation. In return for access to uranium deposits near the town of Shiprock, New Mexico and with risk-free contracts with the AEC, Kerr-McGee gave jobs to 100 Diné in underground mining operations, but at lower wages than usual. It was said the company further cut costs by lowering worker safety measures. Ventilation in the mines was inadequate, and radiation in one of the shafts reached 90 times the permitted level. By 1979, the easily recoverable uranium deposits were being exhausted, and AEC was phasing out the program. The Shiprock uranium mill was closed in 1980, and the company left 71 acres of uranium mill tailings. The huge tailings pile was located only 60 feet from the San Juan River, the water source for Shiprock and downstream communities. Of the 150 miners who worked underground during the 18 years of Shiprock mine and mill operations, 18 died of radiation-induced lung cancer by 1975, another 20 died by 1980, and an additional 95 had respiratory ailments, cancer, and saw deformities among them and their children. By 1980, the Navajo Tribal Council had allowed 42 uranium mines and 7 uranium mills onto or adjacent to their lands.[28]

Tuba City, Arizona

Tuba City, Arizona which previously had a uranium processing mill on the Navajo reservation, was left with mill tailings piles and adverse human and ecological effects like Shiprock. Several persons including babies were reported to have traces of uranium in their blood, and infections developed in those who showered with local water.[29]

Red Mesa, Arizona.

In this area, uranium continued to seep into groundwater from old mines, and persons that used this water were said to have experienced shorter life spans and reduced IQ.[29]

Church Rock, New Mexico

In the 1980s, the Kerr-McGee mine at Church Rock, New Mexico was discharging 80,000 gallons per day of radioactive dewatering fluid from its primary mine shaft into local and downstream potable water supplies. This discharge started as early as the 1960s. In July 1979, the United Nuclear uranium mill also at Church Rock had a break in its tailings pond dam with the release of 3 million gallons of highly radioactive water into the Rio Puerco River. This break occurred shortly after the Three Mile Island accident in Pennsylvania. United Nuclear reportedly knew of cracks in the dam two months previous as reported by the Corps of Engineers, but took no action. The water source serving 1700 Diné was contaminated, and livestock died. The company was said to react slowly, finally making a minimum out-of-court settlement a year later. Church Rock was said to be held hostage because nearly 80% of its economy was derived from the uranium industry. This is now a Superfund site. The EPA wanted people to relocate to Gallup, but many resisted.[28,30]

Uranium Mines in Grand Canyon Area

The Grand Canyon Trust has focused on the Grand Canyon area where there are still hundreds of abandoned uranium mines which hopefully can be cleaned up. Uranium is now found in U.S. homes and in human tissue. U-238 and U-235 have extremely long half-lives. Contamination of groundwaters was of great concern. On the South Rim of the Grand Canyon, about one mile east, there is the abandoned Orphan mine which was closed in 1970 and involved an ongoing cleanup costing $15 million to date. There is also the Canyon mine located about 10 miles from the South Rim. Some 30 million gallons have been pumped out of this mine so far, containing uranium in this water. Figure 3.4 shows uranium mine claims in and around Grand Canyon National Park, and the former Canyon Mine. Figure 3.5 shows the many Native American tribes/nations along and near the Colorado River and the Park. The Havasupai tribe has been exposed to groundwater contamination by the Canyon mine. Twenty-one Native American tribes/nations have called for the stoppage of uranium mining in the Grand Canyon area requesting that a temporary mining ban be made permanent.[31]

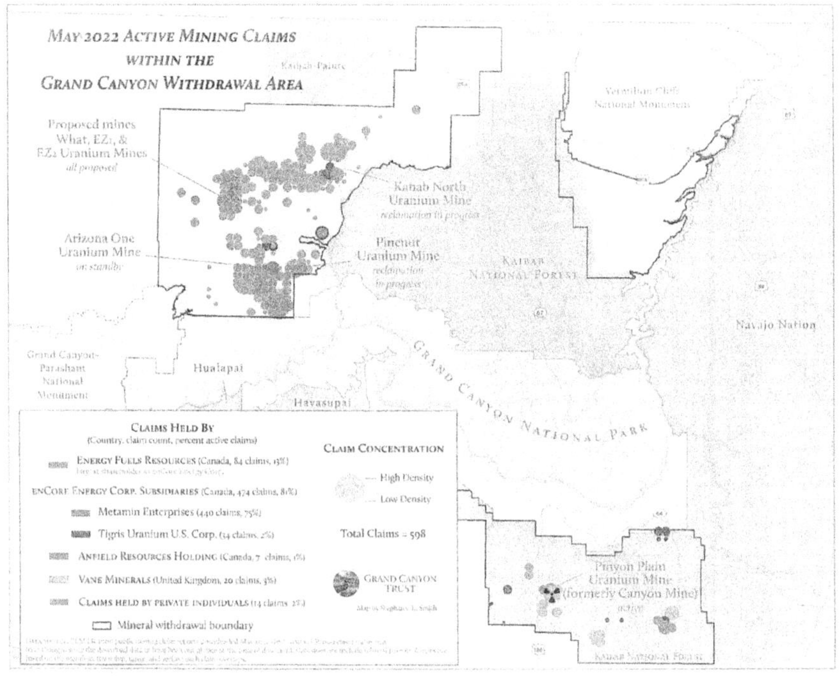

Figure 3.4 Uranium Mine Claims in Grand Canyon Area and Former Canyon Mine

Credit: S. Smith, Grand Canyon Trust

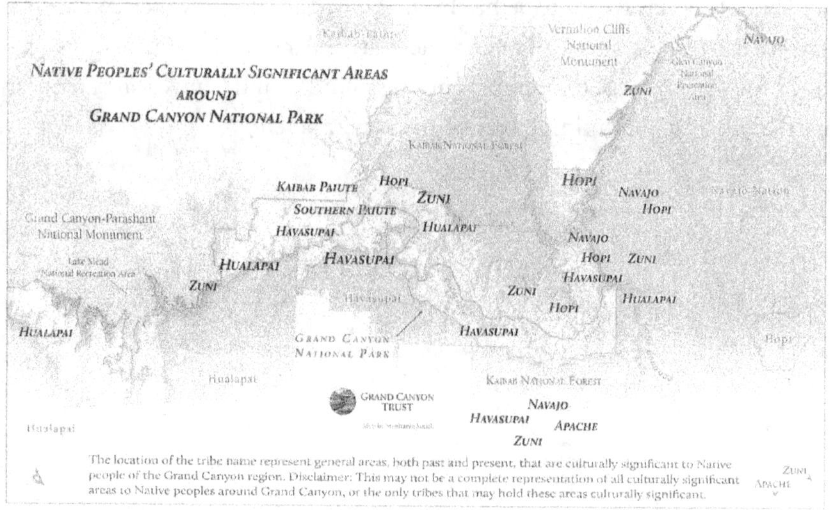

Figure 3.5 Native American Tribes Around Grand Canyon Area
Credit: S. Smith, Grand Canyon Trust

Laguna Reservation, New Mexico Uranium Strip Mining

On the Laguna pueblo reservation in southeast New Mexico, some 7,000 acres of the total of 418,000 acres on the reservation, were leased to the Anaconda Corporation around 1952 for a uranium ore surface stripping operation. In return, the tribe would receive jobs and royalty revenue. Through 1981, 93% of the Anaconda labor force came from the pueblo, the number peaking at 650 employees in 1979. Through the 1970's, unemployment was lowered to 25%. The bubble burst in 1981, when Anaconda suddenly pulled up stakes and left a gaping crater, and piles of radioactive slag. There was little provision in the lease agreement for cleanup and remediation. Water then became a major problem at the pueblo. The Rio Pagute River, which provided irrigation for flourishing agriculture, traversed through the abandoned Anaconda remains, contaminating Laguna's water supply. As early as 1973, the EPA said the strip-mining operation was adversely impacting the surface water supply. The pueblo thereafter converted to groundwater use, but by 1978, all the pueblo's water sources were found to contain radioactivity, as also the soils and new buildings on the pueblo.[28]

4

URANIUM MILLING

From the mines, the uranium is conveyed to a nearby uranium mill as ore material or uranium in solution form. Ores mined in open pit or underground mines are crushed and leached in a uranium mill which is a chemical plant designed to extract uranium from the ore. In most cases, sulfuric acid is the leaching agent, but alkaline leaching is also used. Not only uranium is recovered, but other constituents such as molybdenum, vanadium, selenium, iron, lead and arsenic must be separated out. The final product from a uranium mill is commonly referred to as "yellow cake" (U_3O_8 with impurities), but which is not always yellow in color. As of 1966, there were 17 uranium mills in the Colorado River Basin in the states of Colorado, Utah, New Mexico and Arizona as shown in Figure 4.1.[15,21,24,25]

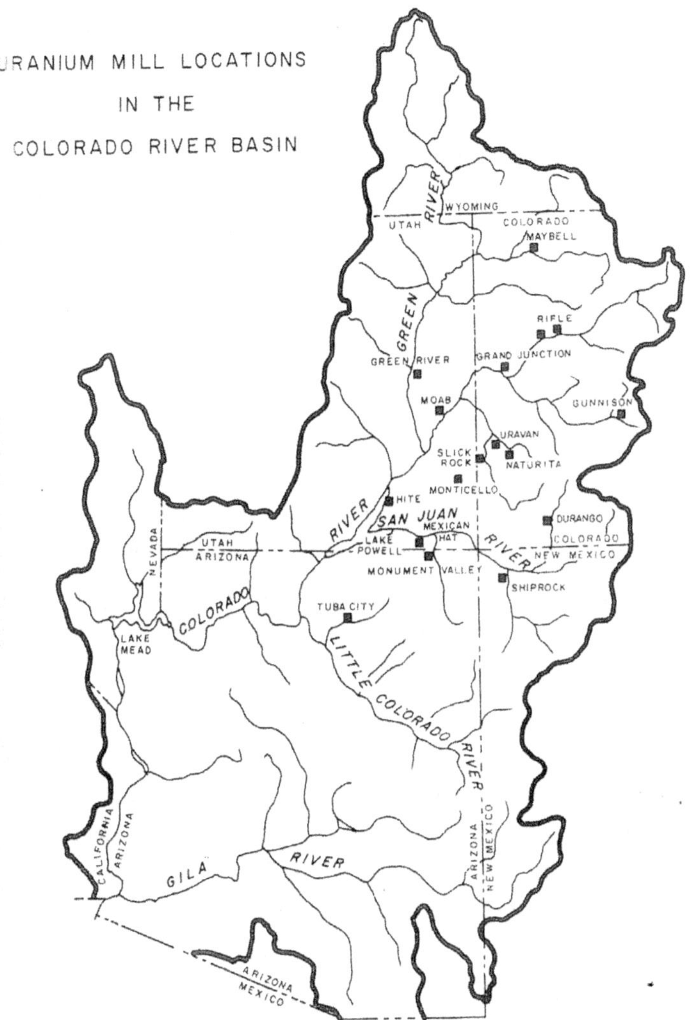

Figure 4.1

Source: USDHEW, Region 8

The major environmental problem with uranium mills was the high radioactivity levels at the mill site and the enormous amounts of liquids and sludges produced as waste product from the uranium mill known as "tailings". The amount of sludge produced is nearly the same as the volume of incoming ores. At an incoming grade of 0.1% uranium in the ore, 99.9% of the incoming material to the mill ends up as sludge which also contains 85% of the initial radioactivity of the ore. Since all the uranium in the ores cannot be extracted, the sludge also contains 5% to 10% of the uranium in the incoming ores. Additionally, the sludge contains heavy metals and other contaminants such as arsenic together with chemicals that were used in processing. Millions of tons of tailings have been generated by uranium mills both in the U.S. and other countries and placed in large ponds or piles, many located next to streams and rivers, where most were abandoned. There were many failures of the dams holding back the semi-liquids in the uranium tailings ponds/piles. Notable dam failures included in 1977 at Grants, New Mexico- a spill of 50,000 tons of sludge and several million liters of contaminated water; and at Church Rock, New Mexico in 1979- a very large spill of more than 1,000 tons of sludge and about 400 million liters of contaminated water.[24,25]

In a tailings pond, the sludge can change to sand which makes it more susceptible to wind dispersion. Radon-22 gas emanates from the tailings piles and has a half-life of only 3.8 days. This seems short, but due to continuous generation of radon from the decay of radium-226 which has a half-life of 1600 years, radon is a long-term hazard, and an agent of lung cancer. Further, because thorium-230 with a half-life of 80,000 years is the parent of radium-226, there is continuous generation of radium-226. Figure 2.1 (given previously) shows the members of the uranium-radium family. It is said the radiation of mill tailings stabilizes after one million years to a level 33 times that of uncontaminated material. Tailings piles are subject to many kinds of erosion, were generally poorly constructed, and seepage from the ponds contaminated groundwater and surface streams. Dried tailings, because of their sandy texture, have been used in construction of homes and are associated with high levels of gamma radiation and radon. In the early years after WW2, mining and milling facilities were not even demolished and simply abandoned. In Canada, uranium mill tailings were often dumped into adjoining waters.[15,24]

The U.S. government belatedly recognized the hazards of uranium mill tailings with the passage of the Uranium Mill Tailings Radiation Control Act of 1978 (UMTRCA). This Act established two programs to protect the public and the environment from uranium mill tailings. Under Title I, DOE was charged with completing surface reclamation at more than 20 *inactive* uranium mill tailings piles. And under Title II, cleanup would occur at uranium recovery facilities that were licensed by the NRC as of 1978, and more or less *active*. Regulations for the protection of groundwater were not included in the 1978 legislation, and only promulgated in January 1995. It is further noted that several mill tailings piles were on the Superfund list. In 1998, DOE testified to Congress it would cost about $2.3 billion (in 1998 dollars) to clean up uranium processing facilities nationwide under UMTRCA.[21,25]

The yellowcake from the mill containing about 0.7% U-235 is next sent to an enrichment plant. The material is changed into gaseous UF_6 and in a gaseous diffusion plant is passed through a membrane filter many times, whereas in a more modern gas centrifuge plant, the U-235 is separated out using large centrifuges. The enriched uranium containing about 3% U-235 is then changed back to solid form. As of 2021, the newer gaseous centrifuge (GC) enrichment plants were operating in six countries. Four new commercial GC plants were under construction, one in France and three in the United States, which indicates continued and/or increased demand for nuclear power plants or weapons production.[24] The focus in this chapter will not be on enrichment plants but on uranium mills. Enrichment plants are briefly described in Chapter 5.

State and Federal Conference on the Matter of Pollution of Interstate Waters of the Colorado River Basin and its Territories, Seven Sessions, 1960 to 1972

In the Colorado River Basin in the Southwest United States was located many of the uranium sources of the Nation. Uranium mining and milling was causing pollution of surface and groundwaters and potential health issues. One area of particular concern was the Animas River in Southwest Colorado into which radioactive waste was being discharged by a uranium mill at Durango, Colorado. The several states of the Colorado River Basin asked that action be taken resulting in the U.S. Surgeon General calling for a Conference under Section 8, Public Law 660, 84th Congress. There were seven sessions of the Conference which initially were held annually. The state conferees were the official participants invited to work with the newly formed Colorado River Basin Water Quality Control Project located in Denver Colorado with its analytical laboratory in Salt Lake City, Utah. The *First Session of the Conference* was held in January 1960 in Phoenix, Arizona, and the *Second Session of the Conference* was held in May 1961 in Las Vegas which set goals and objectives for the Project. The federal agency heading the Project was the U.S. Public Health Service (USPHS) under the Department of Health Education and Welfare. In essence, this conference was a chronology of coping with the massive problem of uranium mill tailings in the Colorado River Basin from 1960 to 1972.[32,33]

At the *Third Session of the Conference* held in Salt Lake City in May 1962, the Project reported elevated levels of radium-226 in streams below two uranium mills- Montezuma Creek downstream of the Monticello, Utah mill and the San Miguel-Dolores River system downstream of the Uravan, Colorado mill. There was also indication of relatively high levels of radium in crops-alfalfa and grasses related to radium in the topsoil. A preliminary technical meeting was held in Santa Fe in February 1963 prior to the Fourth Session. The USPHS in the *Fourth Session of the Conference* held in San Diego in May 1963 recommended that more serious consideration should be given to the ultimate disposal of uranium mill tailings, and the question should be answered as to their suitability for landfill operations and highway subgrade material. The conferees expressed concern of the ultimate disposal of tailings from the Durango, Colorado and Shiprock, New Mexico mills because there seemed to be no definite responsibility for this problem.[34,35,36]

The *Fifth Session of the Conference* was held in May 1964 in Las Vegas. It was said abandoned tailings contained about 1,900 curies of radium-226 and about 1,400 curies of thorium. Two mill tailings sites located very close to rivers containing relatively large quantities of radioactivity, were those at Durango and Naturita, Colorado. Apparently, the AEC had not yet determined who had jurisdiction over abandoned tailings when their average uranium content was below .05%. The AEC also said use of tailings as subgrade material for highway construction was still a possibility. The Conference Chairman said it was imperative the problem be resolved, and suggested the Project prepare a report on this matter including an inventory on every tailings pond or pile in the Basin to be ready by the next Conference Session.[37]

The Project Report on Uranium Mill Tailings Piles in the Colorado River Basin

This report was made available in March 1966 and distributed to the state conferees, the AEC and others.[25] It included a map of uranium mill locations in the Colorado River Basin. At the end of 1961, the uranium mills had an aggregate design receiving capacity of 6,650 tons of uranium ore per day. The accumulation of tailings piles ranged in size from several thousand to several million tons of material at various locations in the Basin. Radium-226 and thorium-230 with very long half-lives are contained in the tailings. It is known that radium in the tailings is leached out and dissolved in flowing water.

Uranium family members are: $U^{238} > Th^{234} > U^{234} > Th^{230} > Ra^{226} > Rn^{222} > Po^{218} > Pb^{214}$

A total of 3 million tons of tailings existed at the seven closed mills containing about 2,000 grams of Ra-226 alone, which represented less than one-fourth of the total amount of uranium mill tailings currently piled in the Basin from all non-operating and operating mills. Control measures for these piles were of serious concern. Estimated costs for control were small compared to the $28.8 million in revenue received in 1964 for the uranium concentrate produced by the operating mills just in Colorado. The transport of radium-containing sediments was documented by the sampling of bottom sediments in Lake Mead which showed radium-226 levels three times higher than background in Basin streams above uranium mill operations. The Project made the following recommendations:

- Covering of tailing piles with dirt or a stabilizer, seeding, and building dikes to prevent erosion to be undertaken without delay.
- Offsite distribution of tailings from both non-operating and operating mills be halted until proper review procedures are in place.
- Discussions and binding agreements to be reached quickly regarding long-term public and private responsibility for adequate maintenance of the tailings piles.[15]

The *Sixth Session of the Conference* was held in June 1967 in Denver. Uranium processing was at its peak in the Basin in the 1960s, and there were 17 uranium mills. In the middle 1950s, one mill alone discharged several hundred tons per day of tailings into a tributary of the Colorado River. There were now- in 1967, inactive tailings piles at each of 10 inactive mill sites in the Basin and tailings piles at 7 active mills. The AEC mill at Monticello, Utah was the only one of the inactive piles that had been covered against water and wind erosion. It was said enforcement of tailings piles and stabilization should rest initially with the states involved. These piles can reach significant heights up to 30 to 40 feet. The state of Colorado cited schedules for covering the piles, and the Conference Chairman did not understand why more attention was not being paid to tailings piles at active Colorado mills. At Gunnison, Colorado there were plans to use tailings in construction of an expanded runway at the municipal airport, contrary to previous Conference recommendations. The conferee for Utah said… "We do not suggest that industry should be expected to assume the costs for remediation over the long run… (since) … it seems this was not the intent when uranium contracts were first established. Furthermore, it is not the type of burden which the states can assume." The federal government would then be primarily responsible for remedial costs.[38]

The *Seventh Session of the Conference* was held in February 1972 in Las Vegas. A draft proposal of model regulations for the states was presented by EPA, Region 8 requiring the stabilization of mill tailings, but the proposal appeared to be more of a generic framework rather than providing specifics. Just prior to Colorado adopting radiation regulations in January 1967, the American Metals Climax mill in Grand Junction, Colorado allowed about one-quarter million tons of tailings to be hauled away by local building contractors for construction fill under and around residential and commercial buildings despite more than ample warning of the inherent risk of using these materials. The EPA said further in the U.S. there were well over 100 million tons of waste tailings, and at the level of 250 picocuries per gram, this represented 22,000 curies of radium, which was a significant potential source of radiation exposure for generations to come. The conferees agreed that a tailings pile regulation could be implemented by July 1, 1973, but that action was not achieved.[39,40]

Dear Sir: Your House is Built on Radioactive Uranium Waste, New York Times, October 31, 1971

The New York Times obtained a draft letter written by the Colorado Department of Health (CDH) said to be going out to 5,000 homeowners in the city of Grand Junction in the following weeks. It was to be sent to properties where radiation levels exceeded the U.S. Surgeon General's guidelines and said:

"An official report on the survey of your property for the presence of uranium mill tailings is enclosed. You will note our study has confirmed the presence of uranium tailings on your property and that the radiation exposure rate is higher than the level at which the U.S. Surgeon General feels corrective action is suggested. We wish to point out to you, in all honesty, that there is little precise scientific information about the long-term health effect of low-level radiation, such as exists in your home. We strongly recommend, however, that you make every effort to lower the radiation exposure level in your home by removing the uranium tailings from your property."

Grand Junction property owners were more than surprised. The danger was coming from a gray, sand-like material which was a waste product from the Grand Junction uranium processing mill no longer operating. In uranium mills, a very large portion of the original radioactivity in the ores stays with the tailings. This material had been carted away from the tailings piles of the mill in large quantities as construction fill for foundations in homes and other construction. In May 1971, the CDH recommended to the City Manager and Chamber of Commerce that real estate sales be restricted until it could be determined that the property was free of tailings. In July, the Board of County Commissioners said if tailings were present, building permits would be granted, but only if this material was removed before construction. Despite all the recommendations made by federal and state governments in 1963 on the radioactivity and health hazards of distributing and using this material for practically any purpose, the Grand Junction situation occurred together with other cases.[41]

The Health Department calculated the lungs of the occupants in 10 percent of these Grand Junction homes were known to have been exposed to the equivalent of more than 553 chest X-rays per year. Another part of the Health Department's letter to the homeowners said "no public funds are presently available to pay the cost of tailings removal (from existing buildings). We are exerting every effort to try to get Federal funds set aside for this purpose to relieve the burden this unfortunate situation has placed on Grand Junction residents." An engineering study prepared for the AEC concluded the cost of removing the tailings beneath the homes in Grand Junction would be very high. For a house valued at $32,000, more than $15,000 of work would be required. And the 5,000 homes in Grand Junction were not all. Many homes in Durango, Colorado- another uranium mill town, had been built on tailings, and 14 other towns in Colorado had the same problem. Estimates for repairing all affected homes in Colorado could run as high as $20 million (tremendously more in today's dollars).

Like Grand Junction and Durango, the city of Salt Lake City, Utah also had a large tailings pile inside its boundaries. The New York Times writer said… "despite assurances from state officials that no tailings were used for construction purposes in Utah, some two years ago, a newsman found many homes and other buildings built on radioactive material. Thousands of other homes in the West may be similarly affected. Whenever there is a big pile of fine, sandy uranium tailings that are free for the taking, it seems, people will find it and use it…". The tailings were an unfortunate legacy of what was left over from our country's search for raw material for atomic bombs. Despite reduced government demand, private companies continued to mine and mill uranium, which after enrichment, was used in commercial nuclear power reactors. The piles of tailings were still growing, albeit at a slower rate.[41]

The Times writer further said these mounds of radioactive sand- more than 90 million tons of it in all, were found at some 30 mills scattered over nine states- New Mexico, Wyoming, Colorado, Utah, Arizona, Oregon, Washington, South Dakota and Texas. The tailings piled outside the mills were freely available over the years, and nobody took adequate measures to prevent people from carting this material away. Colorado was said to license and restrict public access to tailings, but there was still the lack of signage and easy access at these sites. The AEC had mostly defined its own regulatory powers. From the beginning, it focused on major radiation hazards such as uranium, thorium, plutonium and strontium, but radium was never on the list that the agency controlled, although its broad legal mandate was to protect the public from unsafe material. When the uranium is extracted from its ore, practically all the radium remains behind in the tailings. Therefore, the tailings are as dangerous as the ore itself. But because the amounts of uranium and thorium left in the tailings were below the level of 0.05% defined as *important* by the AEC, the AEC had not been responsive to tailings laden with radium and other radionucleotides.[41]

Radium slowly decays into a radioactive gas called radon (Rn), which then decays to radon daughters. American uranium miners breathing this radon were very susceptible to lung cancer. Some believed that tailings were safe as fill under roads and airport runways, which are in the open with ample ventilation. But to have it within an enclosure as a home, the radioactive material is another matter. The radium in the foundation fill produces radon gas which seeps through the concrete slab and collects inside the house, being breathed in by its occupants and increasing the risk of lung cancer. Also, radon daughters failing to get through the basement slab, still emit gamma rays, which can penetrate the concrete and particularly impact children playing on the floors. This radiation takes decades to manifest itself in adults but is much faster with children.[41]

AEC Monticello, Utah Uranium Mill and Grand Junction, Colorado Uranium Mill, Lack of Enforcement and Remediation, New York Times, October 31, 1971

In the first ten years of the atomic energy program, many mills simply discharged tailings and radioactive liquids into the nearest waterway. The AEC belatedly asked the mills to keep the amount of radioactive material dumped into the streams down to permissible levels but did nothing to pay for safety measures or enforcing this measure. The AEC knew the tailings piles had become a large burden because it had a mill of its own at Monticello, Utah. As a demonstration project, the AEC flattened its pile(s), covered it with topsoil, and grassed it over at a cost of more than $300,000, but even this project was temporary and adequate only for a period of 20 years or so. The piles would require perpetual maintenance because they would be dangerously radioactive for an incredible 10,000 years, or so. Because of the very high costs both short and long range, the states did not expect that other tailings piles would be covered by the owners voluntarily, and they were aware that unless responsibility for the tailings were assigned quickly, the owners and contractors would use various legal procedures for the states to inherit the entire lot by default.[41]

The New York Times writer said the AEC in 1964, which had control over all the piles, could by law require the mills to remediate the piles, and AEC promised the states that no mill license would be permitted to terminate without a complete review of the tailings problem. But two years later in 1966, the AEC decided that when a mill owner's license expired, further control of the tailings was not required. The chief AEC legal counsel at that time told a Senate subcommittee that any quantities of uranium considered *"unimportant"* were too low in uranium content to impose restrictions. The effect of this action was to remove uranium mill tailings from any control at all. The AEC concluded that uranium tailings piles posed no hazard to the environment, short-term or long-term and therefore, in 1966, some 90 million tons of radioactive sand were no longer the responsibility of the AEC. Apart from Colorado, where the state took some control, the piles would have to wait until somebody else would step in with control measures. By that time, it would be too late.[41]

1966 was also the year when two individuals from Colorado Department of Health and the USPHS who were in Grand Junction observed trucks unloading fill into an excavation, and the fill was identified as uranium tailings. They questioned the truck drivers and determined much of the sandy fill used in the area was uranium tailings taken from the Climax Uranium Company mill in downtown Grand Junction. (Looking back with what was going on, it is difficult to understand that the AEC, the Climax Uranium Company, and others could have permitted such a practice to happen, and for so long). But it also happened elsewhere. In 1963, Dr. Arthur Warner, the County Medical Director in Durango, Colorado - wrote to the AEC Regional Director- Dr. Donald Walker with serious concerns regarding the use of tailings in the construction of small buildings. Dr. Walker admitted he did not respond in writing but sent to Warner a copy of an AEC letter on the subject- which was not found. In 1964, Mr. Page Edwards, manager of the Durango mill asked the AEC about the use of tailings for construction purposes. The AEC reply was… "tailings… are not subject to the AEC licensing requirements." It seems that serious notice was not taken until 1971 when hundreds and perhaps thousands of homes in Grand Junction and Durango, Colorado were found to have been built on uranium mill tailings.[41]

The Times writer intimated the AEC was keeping this radiation problem from becoming a public issue. Around 1971, the Mesa County (Grand Junction) School Board knew that 15 of its schools had been built on tailings. The AEC at this time was still standing on its position that tailings did not fall under its jurisdiction relying on the 1966 opinion of its legal counsel. A curious position was taken by U.S. Representative Wayne Aspinall from Grand Junction who was on one of the committees for Atomic Energy. He said the costs of tailings removal were too great and government treasuries are too limited for this problem. Governor Love of Colorado said he felt the responsibility in Grand Junction rested with the federal government, specifically the AEC. In 1971, the uranium mill tailings situation was still in chaos.[41]

Phase I Study of Inactive Uranium Mill Sites and Tailings Piles, AEC/EPA, October 1974

This report, consisting of six major sections, was prepared in response to a Congressional Hearing by the Subcommittee on Raw Materials of the Joint Committee on Atomic Energy held on March 12, 1974. The hearing dealt with the Vitro mill tailings site in Salt Lake City, but the hearing proposed a comprehensive study be made of all such inactive uranium piles which resulted in the Phase I report. In May 1974, 21 inactive uranium mill and tailings sites were visited by the AEC, the EPA and the states to determine site conditions, need for corrective action, ownership, proximity to populated areas, and prospects for increased population at the site. Of the 21 sites, 10 had no license. A preliminary report would be prepared for each mill site to determine if a detailed engineering assessment was necessary as Phase II. The second phase of the study would include evaluation of the problems and examination of alternative solutions, preparation of cost estimates, and detailed plans and specifications for alternative remedial action.[42]

Section II of the Report gave existing conditions at the 21 mill sites. Conditions at the Vitro Salt Lake City were completely unsatisfactory, and a Phase II study was to consider waste removal offsite. No stabilization was seen at six of the sites. Pond/pile stabilization at Tuba City, Arizona, Shiprock, New Mexico and at the Colorado mills was found unsatisfactory. Tailings had been taken for use from the sites at Grand Junction, Durango and Old Rifle in Colorado; Lowman in Idaho; Shiprock, New Mexico; and Salt Lake City, Utah. The EPA said radon gas emanating from a tailings pile may persist for half a mile, and there might be movement of radium and soluble salts into groundwater. Where housing and other structures remained from the milling operations, they had frequently been reused for commercial and community purposes. At Salt Lake City, a sewage disposal plant was operational on the site. Construction of an auto raceway began in the middle of a tailings pile but had stopped.

Section III of the Report provided recommendations for Phase II studies. Moving the piles should not be done without compelling reason. Stabilization and decontamination were advised at nearly all the sites and long-term control was specified. Phase II sites were six and included: Tuba City, Arizona; Durango, Grand Junction, Old and New Rifle all in Colorado; and Salt Lake City, Utah.[42]

Section IV of the Report dealt with legislative authority and funding. For the Grand Junction case, the AEC was authorized to provide up to 75% of the cost with the federal share not to exceed $5 million. The Phase II study was estimated to have a cost of about $1 million and require two years.

Section V of the Report gave the background and nature of the problem. Stabilization projects had been undertaken to minimize hazards, but none through 1974 were completely successful. On the radiological side, exposure to humans could include:

- Inhalation of windblown tailings, with sources of Th-230, Ra-226, Ra-222 and its progeny.

- External whole-body gamma exposure from tailings primarily Rn-226 and its progeny, and surface contamination around the piles.
- Ingestion by man of surface and groundwater with Ra-226 leached from the piles, by flooding, soil and food contamination.

On the chemical side, waste substances remaining in the tailings could include sulfuric acid, sulfates, carbonates, chlorides, nitrates, ammonia, potassium permanganate, copper sulfate, cyanides, flocculants, phosphates, various amines, alcohols, kerosene, fuel oil, iron, copper, vanadium, molybdenum, lead, fluoride and others.[42]

Section VI of the Report dealt with Phase II studies for which the AEC and the EPA did not have the capability to perform, so this work would be contracted out, with AEC and EPA laboratories assisting with radiological analyses. The work would start at the Salt Lake City site, and all Phase II work was expected to be finished by June 30, 1976. Corrective action was not expected to start before Calendar Year 1977 and require four years to complete (which was later found to be optimistic). The average cost in Grand Junction to remove tailings from under and around dwellings was $9,000 for each household.[42]

Guidelines for Cleanup of Uranium Tailings from Inactive Mills, DOE, Office of Scientific and Technical Information, Oak Ridge Laboratory, January 1975

This report suggested that current methods for perpetual care and isolation of the large areas covered by tailings piles at inactive mills may be inadequate for minimizing human exposure and it gave proposed criteria for remedial action. The most hazardous components of the tailings were Ra-226 and Th-230 whose long half-lives required consideration of continuous occupancy in future times. The average operating lifetime of a uranium mill was only about 12 to 15 years. Consequently, 24 of the original mills were now inactive. Two approaches for remedial action on tailings piles were in-situ stabilization and removal of the materials. In-situ stabilization may be expected to minimize all contamination except radon emissions. It is much less costly than removal of the waste material, but that land must remain restricted.[43] The questions to be answered for stabilization included:

- What degree of public protection can be expected?
- How long will this method produce effective public protection?
- To what degree must access remain restricted?

Removal of the tailings may be practical if another more remote location is available for the burial of the waste material, or if the tailings represented a recoverable mineral resource. The question was when may the former site be considered clean enough for unrestricted access? A residual contamination level of 0.5 picocuries Ra-226/gram in soil above natural background was not expected to be associated with any specific health hazard.

In-situ stabilization consists of covering the contaminated area with a layer of contamination-free material. Stabilization is regarded as only a temporary and a partial solution to the problems associated with inactive mill sites. Periodic inspection and maintenance of the covering material and security fences are required. Earth covering tends to raise the local water table, and if this occurs, groundwater contamination by the covered pile is possible. In general, soil covering less than three meters offers only minor reduction in the emanation of radon. The prime advantage of stabilization is economic.[43]

Uranium Mill Tailings Radiation Control Act Enacted by Congress, November 1978

The Act, otherwise known as **UMTRCA** authorized the EPA to set environmental protection standards for the stabilization, restoration and disposal of uranium mill wastes. It called for safe and environmentally sound disposal, long-term stabilization and control of uranium mill tailings.

Title I of the Act required the EPA to set environmental standards consistent with the Resource Conservation and Recovery Act (RCRA); DOE to implement EPA standards and provide perpetual care for some sites; and the Nuclear Regulatory Agency (NRC) to review cleanup and license sites to the states or the DOE for perpetual care. Title I of the Act designated 22 inactive uranium mill sites for remediation, resulting in the containment of 40 million cubic yards of low-level radioactive materials in 19 Title I waste holding cells. The radioactive materials would be encapsulated in disposal cells approved by the NRC.

Under Title I there were both designated disposal and/or processing sites, i.e., 19 disposal sites and 11 processing sites. Title I disposal sites were to be managed by the DOE, specifically their Office of Legacy Management. 12 disposal sites and 9 processing sites were in the Colorado River Basin. More could be added to this list later. A portion of the cell at Legacy Management's Grand Junction disposal site was left open to receive additional contaminated materials until filled or 2013, whichever came first. This deadline was later extended to 2031.[44]

Title II of the Act identified Title II disposal sites which were uranium ore processing sites active on or after November 1978. The Office of Legacy Management (LM) under DOE as of 1978 managed 6 Title II disposal sites but only one was in the Colorado River Basin. The site owner at these sites to terminate his license must have an NRC approved plan of reclamation of all onsite radioactive waste remaining from process operations and must also provide full funding for inspections and ongoing and long-term maintenance and monitoring. DOE then accepts title to the site for long-term custody and care by Legacy Management. LM may ultimately manage up to 27 Title II sites. A large section of the Act dealt with state cooperative agreements for the state to perform permanent remedial action. DOE would pay 90% of the remedial action taken, and the state the remainder. If Indian lands were involved, DOE would pay 100%. Beginning January 1, 1980, DOE must submit to Congress an annual report showing actions taken, health, safety and environmental hazards remaining, together with recommendations for future actions, and identification of all responsible parties.[44]

Uranium Mill Tailings: Cleanup Continues, But Future Costs are Uncertain, Government Accounting Office Report to Congress, GAO/RCED-96-37, December 1995

Seventeen years have now passed since UMTRCA. GAO reviewed the costs and status of DOE's program for cleanup of uranium mill tailings under the UMTRCA Congressional directives of 1978. The initial cleanup date was March 1990, but this deadline was extended twice. Since legislative authority expired in September 1996, this report was submitted toward reauthorizing the program.[45] The 1978 Act under Title I directed DOE to clean up sites that were already inactive in 1978, aka Title I sites. Title II of the Act covered the cleanup of sites that were to be cleaned up mostly at DOE expense, but affected states were to contribute 10% of the costs of remedial actions. Title II sites were to be cleaned up mostly at the expense of the private companies that owned and operated them, and then turned them over to the federal government or state for long-term custody. Before a Title II site was turned over, the NRC was responsible for creating financial arrangements with the owner/operators to provide sufficient funds to cover the costs of necessary long-term monitoring and maintenance.

DOE after studying Title I sites requiring cleanup, estimated the cost of cleanup of both surface and groundwater contamination from uranium mill tailings around 2014 to be more than $2.4 billion. The cost of surface cleanup to date totaled about $2 billion. Because of a delay in the issuance of EPA's final groundwater standards, DOE postponed the start of groundwater cleanup until 1991. Unfortunately, agreements with the states and tribes had not been reached as to cleanup strategies and the state's financial support. Questions remained as to whether the affected states would provide their 10% share of groundwater cleanup expenses. Further, assumptions that underlie NRC's minimum charges for owners and/or operators of Title II sites for long-term monitoring and surveillance had not been reviewed and updated and did not include ongoing maintenance at each site.[45]

Toxic Legacy of Uranium Haunts Proposed Colorado Mill, Denver Post, 2016

Colorado in 2016 was nearing the possible approval of a new uranium mill in the Paradox Valley of uranium-rich Southwest Colorado named the Piñon Ridge mill proposed by Energy Fuels Resources Corporation- a Canadian company. The federal government and Colorado continued to deal with the immensely expensive and never-ending mess left by earlier uranium mills. Nearly 20 million tons of radioactive tailings sat at inactive mine sites in Colorado where they must be monitored in perpetuity. The state's latest regulations were said to call for the monitoring of new mill wastes for 1,000 years. Despite scarcity of data, The Denver Post found the cleanup costs of uranium mills in Colorado have ranged from $50 million to $504 million per mill. Nevertheless, the state of Colorado was requiring that owners of the new proposed Piñon Ridge mill would need to pay only $12 million in bonds for future cleanup- a totally inadequate guarantee.[46] The license for this mill was eventually denied.[48] Other developments occurring in Colorado were:

The Cotter Corporation mill near Canñon City. Was the last mill to operate in Colorado. It was decommissioned after nearly 100 violations in the past decade, and in 1984 was declared a Superfund site. Cotter and the state argued over the cost and scope of cleanup, which the state said will cost more than $28 million.

The Uravan mill. Was declared a Superfund site with a $120 million cleanup program recently completed, 20 years after the mill was shut down.

The Maybell mill. Operated between 1957 and 1964, was responsible for 2.6 million tons of tailings spread out over 80 acres. Umetco dismantled the Maybell mill and began tailings stabilization. DOE performed surface cleanup from 1995 to 1998 and placed 3.5 million tons of contaminated material into a 66-acre disposal cell with costs of $63.5 million.[47]

The Naturita mill. 138 acres of the original land were contaminated as also the groundwater beneath the site. DOE from 1993 to 1997 removed 800,000 cubic yards of contaminated material and placed it in a disposal site near Uravan. Contamination was left in place on 22 acres because radiation was so high that workers would have been at risk. Cleanup costs were $86.3 million.

The Old Rifle mill. Reactivated by Union Carbide in 1942 for uranium recovery which continued to 1957. One quarter million tons of tailings were moved to the New Rifle mill which operated from 1958 and closed in 1984. Tailings at the new site were stacked 33 feet high. The groundwater under both mills was contaminated which found its way to the Colorado River. DOE cleaned up the tailings and other hazardous materials from 1988 to 1996 relocating 3.8 million cubic yards of radioactive materials to a cell six miles north of Rifle. The cost of cleanup was $119.1 million.

The East and West Slick Rock mills. The tailings left behind created contaminated groundwater plumes of 19 acres at one site and 92 acres at the other. DOE did surface cleanup from 1995 to 1999 moving 857,000 cubic yards of radioactive tailings into a disposal cell at the cost of $50.4 million.[47]

Uranium Mill Tailings Management Plan, CDPHE, Updated June 2019

This plan was intended to be used in managing Title I uranium mill tailings in Colorado. DOE identified nine inactive or abandoned sites in Colorado. This report focused mainly on what happened at the former Climax mill located in the heart of Grand Junction, Colorado. Over the state, approximately 15 million cubic yards of tailings had been moved to controlled disposal sites, but it was suspected another half million cubic yards had yet to be officially found.[49]

Because of availability at little or no cost, the misuse and dispersal of radioactive tailings were widespread. Examples of its use were in soil attenuation, concrete mix, bedding for concrete and for utilities, in stucco and in brick production. In time, new uranium tailings will be discovered, and disturbance of known deposits will occur. Construction close to these deposits will increase public exposure to gamma and radon exposure with ensuing health effects. When these deposits are in closed areas, radon can build up to high levels with greater occurrence of lung cancer. The Public Health Service in the 1950s showed elevated radon levels in uranium mines causing higher cases of lung cancer. At a preparatory meeting with the Colorado River Basin state conferees held in February 1963 in Santa Fe, a USPHS officer warned these tailings should not be used for any construction purposes whatsoever. Many decades have passed since these warnings.[36,49]

In Grand Junction from 1972 to 1987, some 594 structures in Mesa County had radioactive materials removed by government contractors. These massive amounts of tailings were moved to disposal cells located distant from the contaminated sites, the one exception being Maybell where in-situ capping was used. These cells were designed to last from 200 to 1,000 years. The Grand Junction Disposal Cell was located at 488 Hwy 50 in Whitewater, Colorado and was left open to receive tailings from *all* UMTRCA facilities until at least 2023. All disposal cells are to be monitored and maintained long-term by the DOE. Over 70,000 properties have been surveyed for the presence of tailings but participation in this program is voluntary, and many tailings remain unaccounted for. Tailings when found are removed and hauled to the Interim Storage Facility (ISF) in the center of Grand Junction, and vehicles and personnel decontaminated. Materials in the ISF are eventually taken to the Whitewater controlled cell. It appears that property owners would bear the cost of excavating, stockpiling and transporting tailings to the ISF or to Whitewater. DOE operates the ISF and the Whitewater disposal cell. The EPA expects gamma radiation to be held below 100 millirem per year for the public, and limits are in place for the radiation workers.[49]

Special Case Studies of Radioactive Cleanup

A close look is made of four uranium processing mills, three of which had ceased operation- the Durango and Uravan mills in Colorado and the Moab mill in Utah which are inactive. The White Mesa uranium/alternative fuels mill in Utah is the one remaining active mill in the Southwest United States.

The Durango, Colorado Uranium Mill. This mill was started in 1942-1943 operated by the Vanadium Corporation of America (VCA) initially to recover uranium from vanadium mill tailings to produce a green sludge which was then shipped to the Grand Junction mill to yield yellowcake. In 1946, the green sludge plant was torn down. The AEC bought the plant during the Cold War and leased it back to VCA who operated the plant as a uranium mill starting in 1949 and ending in March 1963 when VCA moved its milling operations to Shiprock, New Mexico with the purchase of the Kerr McGee mill. VCA merged into the Foote Minerals Company in 1967.

The Durango mill received an average of around 500 tons of ore which were crushed and pulverized into powder. Chemicals were added to the powder including about 65,000 pounds per day of sulfuric acid together with salt and soda ash, and 325 pounds of tributyl phosphate and 370 gallons of kerosene. One ton of ore yielded about 6 pounds of uranium, with the rest ending up as waste. The Durango mill was said to release about 997,000 pounds of tailings each day, containing unreclaimable uranium along with lead, polonium, bismuth, thorium, radium, arsenic and other material from the ore. The leaching chemicals were part of the mix. The waste was directed into various ponds which were frequently breached plus two liquid flows of about 340 gallons per minute (gpm), which went directly to the Animas River. The river, besides radioactivity, had high levels of zinc, arsenic, aluminum lead and other toxic metals both from upstream mining activities and the Durango mill.[50,51]

The Durango mill tailing ponds were located directly next to the Animas River which was an interstate stream and used downstream for drinking water, crop irrigation and livestock watering. DOE carried out an extensive cleanup program of the Durango mill site in the 1980s with removal of 2.5 million cubic yards of radioactive material across the Animas River to a burial site in Bodo Canyon 3.5 miles SW of Durango. The cleanup was completed in 1991 after which the ownership reverted to the city of Durango and the Animas and La Plata Conservancy District with long-term maintenance and management of the Bodo Canyon burial site under DOE. Cost was said to be $500 million.[50,52,53]

Durango was no stranger to what was left behind from the town's legacy with uranium mining and milling. Over the years, people freely used the tailings in construction around the town. One could just drive their truck to the waste pile and take a load. As Durango grew, the easiest thing to do was to use these tailings, and people did not understand the real danger. This sand material was used for the foundation of buildings and homes, and for driveways and roads. It was even used as a substitute for sand in gardens and sandboxes. In the 1980s, DOE estimated some 122,000 cubic yards of radio-active tailings had been used in and around Durango homes, businesses, public buildings, roads and parks, and it would take years and millions of dollars to remove all this material. Federal government officials went up and down Durango streets looking for hot spots, and eventually the high-risk spots were removed- an estimated 915 properties were impacted.

However, certain properties were missed, and in 1997, the CDPHE announced even more hot spots were found which unsettled the city of Durango. Contamination potentially existed on about 115 more properties. While most were cleaned up, some escaped that cleanup program. State officials suspected new hot spots were passed over due to tailings being relocated, properties only partially being cleaned up originally, or the homeowner at the time refusing to participate. Homeowners were not required to do testing. However, if a seller was aware of an issue, they were legally bound to share that information with the buyers. Unlike the 1980s when the government paid for cleanup costs, this time around, the cost of cleanup would fall on the property owner.[50,51,52,53]

Uravan, Colorado Uranium Mill. The Uravan mill was a major uranium processing facility in the middle of the uranium-rich belt of southwest Colorado. It was about 90 miles southwest of Grand Junction and on the banks of the San Miguel River which drained into the Colorado River. It began as a radium recovery plant in 1915 and expanded to include vanadium recovery in 1935. The town of Uravan was established in 1935 to house mill and mine workers and their families. The plant operated as a uranium processing facility from the 1940s to 1984, the operator since 1964 being Umetco Minerals Corporation, a subsidiary of Union Carbide. Until the mill was shut down, it processed over 10 million tons of uranium-vanadium ore, and produced more than 10 million tons of tailings, 38 million gallons of waste residue, and other milling wastes. In 1986, the mill site was declared a Superfund site. The complex mixture of wastes included liquid waste from uranium processing called raffinate, radioactive tailings, lead, arsenic, cadmium, vanadium, thorium and various salts.

Soil surrounding the site and groundwater under the Uravan mill contained radionuclides and heavy metals. The air at the site contained elevated levels of radon gas. Cleanup of the 680-acre site was started in 1987 and lasted 20 years until 2007 costing more than $120 million. There were more than 13 million cubic yards of mill tailings, evaporation pond precipitates, water treatment sludges, contaminated soil and debris resulting from more than 50 mill structures on the site. Some 1.5 million cubic yards of waste previously deposited directly next to the San Miguel River were moved to a secure disposal area. The wastes were collected and disposed of in four on-site disposal cells that were closed, capped with soil and revegetated. These cells also contained waste from a nearby abandoned Gateway, Colorado mill, and tailings from the Naturita mill site. More than 380 million gallons of contaminated liquid from the seepage containment and groundwater extraction systems were treated during cleanup. Eventually, the entire town of Uravan was evacuated and demolished because of high levels of radioactivity throughout. The EPA claimed that wastes have been removed, the area has been restored, and threats to the San Miguel River eliminated. The site will be transferred to DOE for long-term management. Plans were to convert portions of the site into a campground, historical mining museum and even a future recreation and wildlife area.[54,55]

Moab, Utah Uranium Mill. The original Moab uranium mill was built on the floodplain of the Colorado River by the UTEX Corporation to provide ore from the Mia Vida Mine to supply the AEC. It was later purchased by the Atlas Minerals Corporation and continued operation until the government stopped buying uranium. Atlas rebuilt the mill to process vanadium-bearing ores. This was the only AEC venture that was expanded during the uranium producing era. Its tailings piles were unlined and overlaid porous geological strata where the leachate went directly into groundwater. It was also the only Title II tailings site in the U.S. which sat over an area that was known to be seismically unstable.

In 1996, reclamation of the Atlas tailings piles was discussed by the government. The EPA preferred the tailings be moved to an alternative site but since relocation was five times more expensive than on-site stabilization, the latter approach was favored. The quantity of tailings at this time was 10 million tons, and there was the need to address groundwater contamination. The federal government estimated the capping proposal to cost $11 to $17 million but the state and county estimates were $36 million or more. In April 1998, a coalition of environmental groups said it would sue Atlas to force removal of the tailings outside of Moab. It said contaminants from the 10.5 million tons of tailings were leaching into the Colorado River just 750 feet away, polluting the river. Meanwhile, in September 1998, Atlas filed for bankruptcy in federal court under Chapter 11 so it could continue to operate and generate funds to pay off its creditors. Later in 1998, Atlas said it intended to reclaim the tailings and had sufficient funds to do so. It would obtain insurance beyond the $19 million cost estimate for onsite reclamation.

In March 1999, it was revealed the Company only had between $9 and $11 million that could be used for in-situ reclamation while the estimated cost was $16 to $19 million. In 1999, the U.S. House of Representatives introduced a bill to transfer responsibility of cleanup of the Atlas tailings to the DOE to move the tailings from the floodplain of the Colorado River which would take an additional $155 million, instead of capping it. Atlas would also be released from all future liability arising from its mill and tailings impoundment. The Nuclear Regulatory Commission (NRC) still hoped that stabilization could be used.

The chairman of the Atlas Mill Reclamation Task Force working with NRC since 1990 described the Atlas situation as only something George Orwell would understand. The Atlas tailings piles stood 110 feet high on 130 acres on the floodplain of the Colorado River bisected by the geologically active Moab Fault next door to the city of Moab and the Visitor's Center for Arches National Park. The total Atlas site was 400 acres. The Atlas tailings piles were leaking alpha radioactive material into the Colorado River at levels 1,300 times above EPA suggested levels, ammonia in the leachate was 1,500 ppm, and several metals were found in high amounts. Even Los Angeles was fearful of leaving these piles in place. Capping the pile in place would have little or no effect on curbing the leachate. Local groups came up with an alternative plan to move the tailings by rail to a site 23 kilometers north of the present site and be buried below grade in a Mancos Shale stratum.

Finally in 2005, DOE announced that the radioactive uranium tailings from Atlas would be moved, predominately by rail, to a holding site at Crescent Junction, Utah about 30 miles from the Colorado River. At the new site, the waste would be buried and covered in a large pit lined with a protective layer to keep the material from seeping into the groundwater. Cleaning up and moving the original pile was estimated by DOE to have a cost of $472 million with a completion date of 2017. Also, active groundwater remediation would be performed at the original site. The estimated amount of tailings was now 16 million tons from the former uranium ore mill that had operated for nearly 30 years. In April 2009, the first tailings started to be moved to an engineered cell constructed near Crescent Junction, Utah. Before transport, the drier sands were mixed with wetter slimes and disked to achieve a desired moisture content. The tailings were then placed into metal containers and hauled to a rail area across a state highway. Two gantry cranes loaded the containers onto a train which ran twice daily to the disposal site. At Crescent Junction, the containers were off-loaded into haul trucks for transport to the disposal cell. Sheepsfoot rollers compacted the tailings in the disposal cell. Details were not given on the disposal cell, but it was very large in size. An interim groundwater remedial action plan was initiated in 2003 for the cleanup site. Eleven extraction wells close to the original tailing piles were installed for groundwater remediation. By 2010, more than 155 million gallons of contaminated groundwater had been extracted including 640,000 pounds of ammonia.

In 2007, DOE reported the relocation of Atlas tailings would take five times longer than initially projected, potentially dragging through 2028, and by July 2008 the cost estimates had greatly increased to between $844 million and $1.1 billion. The eight-million-ton mark (halfway) was reached in January 2016, but budget cuts and the need to excavate more room in the Crescent Junction disposal cell would likely slow down the remedial work. In May 2016 it was announced the completion date will not be until 2034 due to budget cuts. In April 2019, 9.5 million tons had been moved. More than 60 years have elapsed since the first studies of uranium mill tailings by the USPHS in the Colorado River Basin. Progress has been slow, and the author believes the total cost for the Moab cleanup and remediation will easily exceed $1 billion.[35,56,57,58,59]

White Mesa Uranium Mill, San Juan County, SE Utah near Blanding. This mill did not appear on scene until 1979 and started up in 1980. In 1997, the mill was run by the International Uranium Corporation (IUC) based in Denver and permitted to receive alternate materials from all over the U.S. and abroad. Since 1987, the mill had been receiving "alternate feed materials" from various FUSRAP sites in the United States, which capacity the AEC and DOE desperately wanted. The Formerly United States Remedial Action Program (FUSRAP) was initiated in 1974 to identify, investigate and, if necessary, clean up or control sites throughout the United States contaminated because of Manhattan Engineer District or early AEC activities. Alternate feed can be material from rare earth mining, trucked to the White Mesa site mainly as sludges from locations shown in Figures 4.2 and 4.3. In 2019, a waste stream was also received from Estonia. White Mesa derives practically all its revenue from the processing of alternate fuels rather than from uranium ores. The incoming material has only 0.3% of the desired concentrate whereas the remainder of 99.7% ends up in slurry tailings ponds.

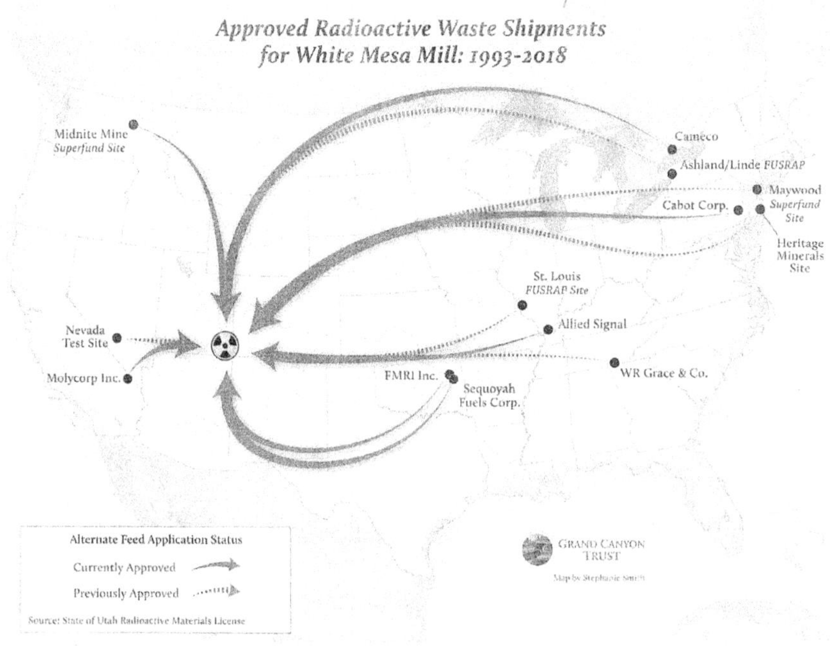

Figure 4.2

Credit: S. Smith, Grand Canyon Trust

Figure 4.3

Credit: S. Smith, Grand Canyon Trust

In May 2000, IUC admitted its tailings ponds capacity at the White Mesa site was insufficient for processing of more alternate feed material. In 2002, the USEPA expressed concern about expansion of alternate feed material at uranium mills and mill tailings impoundments and asked for a thorough review, but receipt of alternate feeds continued at White Mesa. In December 2006, Denison Mines assumed operations from IUC and ran the mill from December 2006 to 2012. In September 2009 it submitted a groundwater permit renewal. In June 2012, Energy Fuel Resources, Inc. (EFRI), a Canadian company, took over the complete assets of Denison Mines including the mill. EFRI not only owns the uranium mill but also Colorado Plateau mines that supply ores to the mill including Canyon Mine near the Grand Canyon's South Rim. In 2018, the state of Utah requested that groundwater compliance limits for the White Mesa uranium mill be relaxed, and this same year the state approved a 10-year renewal of the mill's license and a five-year renewal of its groundwater discharge permit.[60,61,62,63,64]

In 1999, chloroform at levels 47 times higher than allowed by the state of Utah was found in a groundwater monitoring well at the site. A "walk" of citizens and environmental groups from Blanding to Moab was held in May 2001 who feared the White Mesa mill would end up in a similar way as the Atlas Moab mill tailings site. In June 2012, radon emissions at the 66-acre Cell 2 exceeded radon limits over the year. White Mesa has operated five tailings ponds comprising a total of 275 acres- cells 1, 2, 3, 4A, and 4B. EFRI has requested that 80 acres of pond be added- cells 5A and 5B. See Figure 4.4.

Figure 4.4 Tailings Ponds at White Mesa, Utah Mill
Credit: Bruce Gordon, Ecoflight

Over the years, Ute Mountain Tribe has engaged the White Mesa plant insisting that the mill posed a serious health threat to their reservation three miles away. The Ute Mountain Tribe and other organizations worried that the mill adversely impacted the water quality of vital springs and posed a long-term threat to the Navajo aquifer, the main source of drinking water for southeast Utah and northern Arizona. They said the older tailings ponds were lined with thin layers of PVC installed in the 1980s with a useful life of 20 years and also lacked modern leak detection systems. The Ute Mountain Tribe contended background conditions in the aquifer should not be adjusted to justify relaxation of compliance limits. In 2014, the Grand Canyon Trust sued EFRI to correct ongoing radon problems at the mill and ensure future resources for the cleanup and reclamation of the mill. Besides Cell 2, radon emissions from Cells 1, 4A and 4B had increased dramatically. A Fact Sheet from Grand Canyon Trust in 2021 said the impoundment cells would need to be double lined with a system for detecting and collecting anything leaking through the first liner from escaping to the environment. Utah also did not require an ample bond to clean up the site if the company failed to do so- only $20 million, which was considered inadequate.[60,61,62,63,64]

The environmental legacy of uranium mining and milling in the southwestern United States has indeed been complex and very costly.

PART THREE

URANIUM ENRICHMENT FACILITIES

5

GASEOUS DIFFUSION AND GAS CENTRIFUGE PLANTS

Uranium enrichment is a key step in transforming natural uranium into nuclear fuel for commercial power and for nuclear weapons production. This sector will be briefly addressed.

Gaseous Diffusion Plants

The Portsmouth, Ohio Gaseous Diffusion Plant.

Portsmouth Ohio produced enriched uranium for the AEC and nuclear weapons facilities, and for Navy nuclear propulsion. In later years, it produced fuel for commercial nuclear reactors. Portsmouth was one of three gaseous diffusion plants in the U.S. The other two were the K-25 plant at Oak Ridge, Tennessee and the Paducah plant in Kentucky. Portsmouth was started up around 1954, and now continues under decontamination and cleanup.

The entire site covered 3,777 acres. The plant was located close to the Scioto and Ohio Rivers.

The primary means of enrichment was the use of gaseous diffusion of uranium hexafluoride to separate the U-235 from the heavier isotope U-238. In the mid-1960s, Portsmouth received material from Paducah at 2.75% U-235 and enriched it further to 4% to 5% for use in commercial power plants. 60 to 70 MGD of cooling water was used in eleven cooling towers, of which 20 MGD evaporated to the atmosphere. Portsmouth ceased operations in May 2001 after it combined operations with Paducah. "Depleted uranium" product absorbed in structures and equipment was scheduled for decontamination starting around 2005 which represented a long-term process. But in 2007, operations were started again at Portsmouth this time as a gaseous centrifuge enrichment installation, this demonstration intended to produce a 19.75% U-235 product. A revised license to operate the demonstration plant was extended to May 2022, at which time the full-scale plant would be started up. There were also plans underway to support the DOE goal of demonstrating an advanced reactor by the late 2020's.

Air monitors at a middle school 2 miles distant from the Portsmouth plant registered the presence of airborne neptunium-237 and americium-241, respectively in 2017 and 2018. In May 2019, a report indicated the presence of enriched uranium and transuranic radionuclides within the school. The school was shut down for the rest of the 2018-2019 school year and businesses, homes, soils and groundwater underwent further testing. DOE said testing showed no radioactivity above background. Nevertheless, a class action lawsuit was filed on behalf of area residents.[65,66,67]

Gas Centrifuge Plants

The Eunice, New Mexico Gas Centrifuge Plant.

As of 2021, the only full-scale operating uranium enrichment plant using gas centrifuge technology was that located 5 miles east of Eunice, New Mexico which was started up in 2010. A gas centrifuge relies on centripetal force to separate out U-235 from U-238. The feed material is gaseous uranium hexa-fluoride. The gas centrifuge process was developed at Oak Ridge, Tennessee in the 1980s to replace gaseous diffusion for uranium separation. A high degree of separation is achieved by using many individual centrifuges in a cascade, that results in higher concentration of U-235 using significantly less energy. The gas centrifuge process at Oak Ridge used more than 1300 gas centrifuges and an additional 700 were operated by DOE at Portsmouth, Ohio. The Ohio project was shut down in favor of the Advanced Vapor Laser Isotope Separation process. Besides Eunice, New Mexico, there are two or three other commercial gas centrifuge plants that have been granted license to operate in the United States.

The Eunice, New Mexico facility has grown to 236 employees with an annual payroll of $23 million. At full capacity, it can provide half of current enriched uranium demand for civilian nuclear power plants in the U.S. The Eunice enrichment plant is in an area known as "nuclear alley" which includes the Waste isolation Pilot Plant (WIPP). The price of enriched uranium for nuclear fuel has continued to rise in recent years.[65,66,67]

PART FOUR

NUCLEAR WEAPONS FACILITIES

6

MAYAK, SIBERIA, RUSSIA PLANT

History and Background

The Mayak plutonium manufacturing facility is in Chelyabinsk Oblast of the south-central Urals in Russia. The term "Oblast" refers to a territorial subdivision of Imperial Russia and is used in several countries. The city of Ozyorsk or Ozersk about 5 miles north of the giant plutonium plant, is a closed city founded in 1947 on the shores of Lake Irtyash established concurrently with the plant. Early on, the town and the plant were known as Chelyabinsk-40 and then Chelyabinsk-65, i.e., the last digits of the postal code. Kyshtym was another nearby town. In the first few years, Ozyorsk existed as a phantom city of 50,000 people. Mayak is located relatively close and about 100 miles south of the large city of Yekaterinburg. An aerial location map is given by Figure 6.1. An author's rendition of important surface features is shown in Figure 6.2.[4,68,69,70]

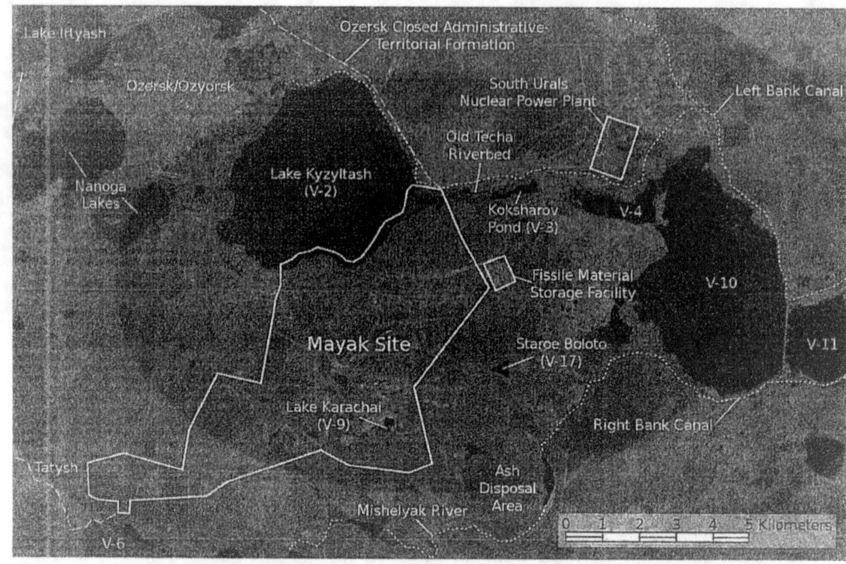

Figure 6.1 Mayak, Siberia Nuclear Facility Satellite Map, April 2010

Credit: Wikimedia Commons and JanRieke

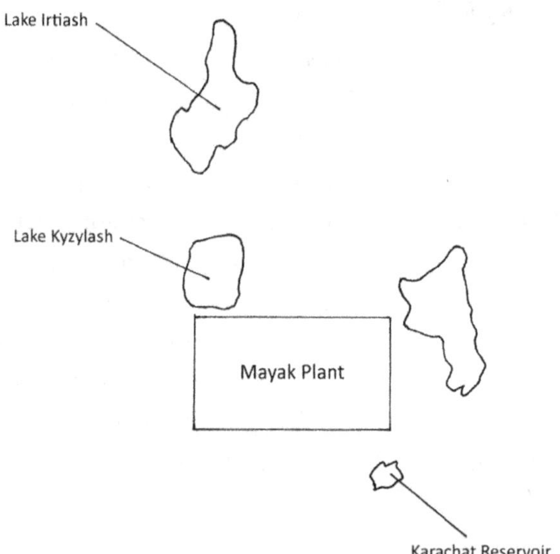

Figure 6.2 Mayak, Siberia Nuclear Facility Site Map,

Author's Sketch

Mayak and Chelyabinsk started up in 1945-1946 compared to its counterpart Hanford, Washington plutonium facility in 1943. In Russia, the Gulag was depended upon as a source of raw labor, and in 1946 prisoners and soldiers built five military garrisons and eleven corrective labor camps to meet the need of forced labor to build the city and the plant. Mayak was the birthplace of the Soviet Union's nuclear program and is believed to be the largest plutonium manufacturing and fabrication facility in Russia. The plant covered an area of 35 square miles.

The first Russian plutonium reactor that was modeled on Hanford was graphite-moderated and water-cooled, but the fuel channels and control rods were vertical in the reactor instead of horizontal, which may have been a contributing factor to the Chernobyl disaster of 1986. It was a top-secret facility. In November 1948, the Soviets had irradiated enough uranium for one plutonium bomb. But then the plutonium unloading platform fell apart, and the workers were instructed to remove fuel cylinders with their bare hands. Soviet safety regulations allowed an annual radiation dosage of 25 roentgens per person, many times the U.S. limit. In the U.S., it was reported there were times when managers asked workers to remove their radiation counters before entering irradiation areas thus evading the radiation limits.[4]

During the first year of operations, 2/3 of Mayak employees received radiation doses around 100 roentgens, while 7% received greater dosages. The Soviets exploded its first atomic bomb just four years after Hiroshima and Nagasaki on August 29, 1949, but their main objective was a hydrogen bomb, substantially helped by atomic bomb spies in the U.S. This bomb was tested in November 1955 in Kazakhstan with an explosive power of 1.6 megatons. "Annushka" was the first of six reactors built at Mayak. Five more became operational between 1950 and 1955. Three of these went operational in one year-1951.

At Mayak, Site A contained the reactors, and Site B the plutonium processing plant. The Tatars and Bashirs who were Muslim religious minorities made up the majority of the surrounding towns and villages but were seldom hired by the Soviets. The first reactor was started up in June 1948. Highly excessive radiation was said to be found everywhere outside the reactor facility, but the Soviets were not concerned. In her book "Plutopia", Kate Brown makes it clear that both the Soviet and American nuclear programs skimped on safety and waste management, but the results at Chelyabinsk were worse, and tens of thousands of the population were doomed to radioactivity exposure to prioritize the production of nuclear material. Chronic radiation syndrome caused many early deaths. At Mayak, this new disease, the sick young people, and early deaths were classified as state secrets. The cumulative impact in Chelyabinsk of repeated accidents combined with years of secrecy, earned this region the unwanted reputation of "the most contaminated place on the planet." [4,68,69,70,71]

Operations

The Soviet Union built the atomic bomb in a very short time, which led to repeated accidents, a lack of concern for safety and health, overwhelming pollution, illness and death. In the early years, the number of workers at Chelyabinsk swelled to 47,000. The chemical plant extracted plutonium from uranium irradiated in the reactors. This was followed by a metallurgical plant to convert the plutonium concentrate to high-purity metallic plutonium for the first Soviet bomb. Workers handled the uranium blocks (or pits) without hand protection and cleaned and removed radioactive waste with rags and buckets. Unknown to them, they were receiving in 10 to 12 minutes more radiation than what was an annual permissible dose of radiation. Soldiers and prisoners working at the plant were the first to die. A predominately female workforce separated plutonium from radioactive uranium by hand. Many of these women died before reaching the age of 30. Radiation sickness was everywhere. [69,70,71]

Mayak dealt with cracked fuel slugs in Reactor A in the beginning, all of which had to be removed from the reactor. In January-February 1949, more than 39,000 irradiated slugs were offloaded by hand, and replaced by new slugs. The first batches of properly irradiated uranium slugs were plunged in water, the protocol calling for 120 days of cooling, but this was reduced to 30 days to speed up the process. High smokestacks were designed to disperse radioactive gaseous emissions over a greater land area, but this also failed to work. These emissions were unfiltered, adding to an already bad situation. At the processing plant- Factory 25- the irradiated slugs were dissolved in nitric acid to distill the resulting toxic cocktail into plutonium. The startup months at Mayak were a disaster with large losses of plutonium. The Russians received processing knowledge from the American spy David Greenglass, a Los Alamos worker who passed technical documents to his Soviet handler.

After separating plutonium from uranium, the plutonium solution was transferred to Area V where it was transformed into metal ingots, and finally into the softball-sized orbs of weapon-grade plutonium for a bomb core. At every step in the plutonium production line, Soviet workers and their supervisors were exposed externally and internally to radioactivity and toxic chemicals. In the first year and a half, 85% of all workers received more than their lifetime permissible dose of radioactivity. In 1949 when two new reactors were still under construction, it was found that Reactor A was leaking badly, sending radioactivity into Lake Kyzylash where a large fishing company harvested and processed tons of whitefish for commercial sale.[69,70,71] After the first operational phase of mass worker overexposure to radiation, accidents slowed down. An explosion occurred in October 1955. In April 1957, six employees were exposed to high dosages of radiation, receiving from 300 to 1,000 rem each, and one woman died. The main source of accidents and radioactive contamination shifted from the reactors to the radioactive complex where plutonium-239 was separated chemically from other isotopes produced in the reactor.

Waste Storage, Discharge to Techa River and Numerous Adjoining Lakes, Impact on Downstream Population, Evacuations

Plutonium production was top priority, and large amounts of radioactive waste had to be disposed of. The immediate solution was to dump the contaminated chemical and biological wastes into nearby lakes and rivers. Although the Mayak facility was far from big cities, Russian, Tatar and Bashir villages dotted the banks of many lakes and rivers. Mayak was picked for its location owing to its abundance of water especially needed for the cooling water systems and chemical processes. The largest of the lakes- *Lake Kyzyltash (Kyzylash)* was a reservoir for cooling water, but the highly contaminated spent waters were dumped back to this same lake. A much smaller lake- *Lake Karachi (Karachat, Karachay)* was simply used as a repository for radioactive and chemical wastes. "A total of 100 million curies of radioactive waste was dumped into the lakes."[4]

At Mayak, liquid wastes were classified as high, medium or low in strength. A Dixie cup of high-level waste in a conference hall could kill everyone in it. Temporary solutions consisted of storing high-level waste in underground storage containers. Low and medium-strength wastes were dumped directly into the adjacent Techa River. The Soviets knew the Americans were doing the same with the Columbia River at Hanford. The Columbia however was fast-moving in comparison to the Techa which was slow-moving and turgid. The Russians knew the Techa River was severely polluted and in late 1949, capacity had run out with the underground storage containers. Ceasing production was out of the question, and in January 1950, orders were given that all the plant's effluent including high-level waste were to be dumped into the Techa River- some 4,300 curies per day. Calling the Techa a river was almost unbelievable since much of the adjoining land was marshland. With spring thaws, the river spread out over the flood plains. From 1949 to 1951, 20% of the river flow was plant effluent.[69,70]

In December 1948, the River Techa had become a radioactive gutter. From 1949 and 1951, 2.7 to 3.2 million curies of radiation were released into the Techa River, upon which tens of thousands of people relied for clean water. Forty-one settlements lined the Techa River downstream of the Mayak facilities. The majority of these were ethnic Muslim Tatars and Bashirs who had been there for centuries. The Techa moves slowly about 150 miles before flowing into the Ob River which eventually reaches the Arctic Sea. The Tatars and Bashirs depend on the Techa River for drinking, irrigating their fields, washing clothes, swimming and fishing. In 1951, river measurements were made which showed dramatically high levels of radiation in Russia's Arctic waters- a thousand miles away, in contributing rivers, and in the soils and the fields of the villagers. Metlino was the first downstream community with a population of 1,200 persons, four miles from the point of waste release from the Mayak plant. With the then current levels of radioactivity, a Metlino resident was experiencing 400 to 600 times normal radiation exposure and could get a lifetime allowable radioactive exposure inside of one week. Worse, the scientists found that the villagers were using pond water and river water for drinking, cooling, bathing, washing of clothes and crop irrigation. Shortly thereafter, many of these residents were evacuated or removed and never heard of again.[69,70]

A sampling program of 1951 involved 19 villages downriver, and the tragedy unfolded. A Commission was formed, and recommendations made, one to cease the dumping of radioactive waste into the Techa River. Canals were dug to divert the waste into Lake Karachay. Wells were proposed for 20 riverside villages but were not completed or became shallow wells. In 1956, there were more evacuations, and the killing and burial of livestock and farm implements. Widespread radiological poisoning of the villagers was found. The radical removal of residents from 16 villages was proposed. In the end, it took more than 10 years to resettle 10 villages away from the Techa. But five riverside villages were never evacuated in the 1950s. Once the large quantities of Mayak waste were stopped from going into the Techa River, they were stored in the nearby 110-acre Lake Karachay, which was more like a bog than a lake, and then the lake dried up with strong winds dispersing highly radioactive dust across a vast territory and affecting thousands of new victims. Highly radioactive material was also seeping into the underlying water table below Lake Karachay. Waste was also stored in at least seven other shallow lakes near the plant. The public was not told of this situation until 1989.

In 1953, the radioactive waste which had been going to the Techa River started to be placed into underground storage tanks which were subsequently covered with concrete. These tanks organized in groups of 20, were positioned in pits 27 feet deep. These tank groups, however, continued to have fission-product decay causing heat emissions. The groups of tanks were designed with water-cooling, ventilation and monitoring systems. Things worked for a while.

Plant managers at Mayak said they would continue to dump radioactive wastes into open reservoirs until at least 2018. Nuclear accidents at Mayak occurred repeatedly but were kept secret. It was later found that the Mayak nuclear processing plant had continued to dump radioactive waste into the Techa River from 2001 to 2014. The Russian government appeared to be waiting until all the victims of radiation died, and then they would not have to pay compensation to the victims. One author said… if nothing else, the 24,000-year half-life of plutonium guarantees job security. The population of Ozyorsk was 91,760 in 2002 decreasing to 82,164 in 2010. The villages of Bashir and Tatar were never evacuated although they were only 200 yards from the evacuated villages. It is said that both Russian and American leaders denied any ill effects to these downstream populations. In 2005, the relatively large village of Muslyumovo, downstream from Ozyorsk, was slated for relocation, but this did not happen because of cost. In 1999, a local doctor estimated that 95% of infants born in Muslyumovo had genetic disorders.[69,70,71]

Massive Explosion of September 27, 1957, Large Scale Evacuations

On this day, the monitoring system measuring the heat level in the high-level waste containment storage tank- Tank No. 14 failed, and the tank exploded. Other waste containers in the area also ruptured. The tank blew and belched up a 169-ton concrete cap that had been buried 25 feet below ground, tossing it 75 feet into the air. Radiation reached 40,000 times accepted levels, and the overhead radioactive cloud was immense. The radioactive dust cloud spread 20 million curies of radioactivity including strontium-90 and cesium-137, over at least 52,000 square kilometers, inhabited by 270,000 people, and traveled 325 kilometers from Mayak. After 10 hours, Ozyorsk was given the order to evacuate after several inches of radioactive dust covered the site. Gray soot was also falling into *Lake Irtiash*, the source of drinking water for the city. Radiation levels in the area were reaching 300 roentgen per hour, but some estimates were as high as 200 roentgens/second. More than 1,000 soldiers received radiation doses more than the permitted norm, 63 of them up to 50 roentgens. Decontamination and cleanup were orders of the day. Whereas the entire site was in a state of emergency, no notification was given to the off-duty workers and citizens of the company town of Ozersk. Another series of radiation monitoring that same day showed that within 100 meters from the center of the exploded tank, gamma radiation reached 100,000 microroentgen/second, and at a distance over one mile, it was between 1,000 and 5,000 microroentgen/second. There was general destruction throughout the affected area. Dosimeters had gone off scale and/or showed radiation levels 40,000 times above acceptable levels.[4,19,69,70,71]

A follow up Soviet commission was concerned with the possibility of new explosions. The one that destroyed Tank No. 14 had also destroyed the water pipes and ventilation systems that were keeping the rest of the waste storage tanks cool. It seemed to be only a matter of time before 19 more explosions could happen. Unfortunately, the corridor in which the explosion had taken place was also the most important one in the entire underground structure, housing the cooling and ventilation pipes for the rest of the banks or groups of tanks. The solution was to drill through three feet of concrete and establish separate cooling systems for each waste holding tank from the outside. People inside the corridors received excessive roentgen exposure although the working shifts were only 20 minutes long.

Three days after the explosion, radioactivity in the area remained unbelievably high, virtually every animal in the village was killed by the authorities, and the residents relocated. Medical doctors in the region were told not to release any illness information caused by radiation. The decision was made to clean up the plant- mostly by hand, rather than replace the contaminated sections of the plant. No fewer than 25,000 soldiers were used in the cleanup, together with many others, and there is no record of the radiation dosages they received. The cleanup took over a year. Fields, pastures, reservoirs and forests were found to be unsuitable for further use. From Moscow, orders came to evacuate the most radiated villages, which was accomplished within two weeks. Several months later, three more villages were evacuated. In the remaining villages, food and animals were closely monitored, but dirty well water continued to be consumed. Following increased radiation sickness, a third evacuation was ordered with 23 more villages shut down. Three years after the accident, there were still hot spots showing up to 60 curies per square mile. The Ozyorsk explosion of 1957, one of the worse nuclear accidents in history, was estimated to have released 50 to 100 tons of high-level radioactive waste over a huge territory in the Urals, but was not made known to the world in full detail until 1990.[4,19,69,70,71]

In the first few weeks after the explosion, as many as 10,000 people were at work at the site, bringing water to the waste tanks or involved in overall decontamination. At the nearby town of Ozersk, alpha radiation levels were 40 times the norm, while beta radiation was 1,200 times the norm. Showers were now mandatory for everyone entering the contaminated zone. Many engineers and skilled workers-some 3,000, several of which were party members, packed their bags and left the city. Of the 20 million curies of radiation released by the explosion, 18 million fell on the plant itself, but information was maintained as a state secret.[4,18,68,69,70]

In October, it was found that the spread of radiological contamination in outlying regions was significant and included the regional capital of Cheliabinsk with a population of 650,000. Decisions of resettlement of adjoining populations were chaotic at best. It was found that cows were the domestic animal most frequently sacrificed. The houses of the evacuees were burned. In February 1958, additional settlements in the contaminated zone were selected for removal and destruction. The residents of seven villages were resettled from the contaminated area, receiving less than equitable compensation for their losses. Most tragic of all were those allowed to stay in their village insufficiently contaminated to justify resettlement. Affected villages had almost every house with a cancer patient in it, and the Muslim relatives of the victims were opposed to autopsy.

Over the period 1950 to 1959, there had been a steep rise in infant mortality, and female workers exposed to radioactivity had children twice as likely to die before they reached the age of 20. When Lake Karachay dried up in 1967, the winds scattered contaminated dust from the lake over many miles away, but this time there was far less panic among the residents than in 1957. It was not until 1994 that Russia's secret nuclear weapons cities finally appeared on the Russian maps including Ozyorsk. When in the mid-1960s, Soviet leaders planned to build a brand-new nuclear power plant near the sleepy town of Chernobyl, Mayak engineers were chosen to furnish the design.[1,4,69,70,71]

In 1989, an American scientist found that Lake Karachay contained up to three times the total radioactivity release from the Chernobyl nuclear accident. He also concluded a person on the shore of Karachay would accumulate within one hour a lethal dose of radioactivity. In 1990, Soviet scientists verified that there was extensive radioactivity up to 60 miles from Mayak. Asanovo Swamp containing 600,000 curies of radioactivity was open to the public, and the world was told of the Techa disaster much later.[69,70,71]

Direct Plutonium Processing Stopped, Reprocessing of Outside Spent Radioactive Materials, Severe Health Issues

Mayak no longer makes plutonium, and now stores and processes weapons-grade plutonium from thousands of decommissioned bombs. Humans living in areas affected by the Mayak plant have developed problems with reproductive functions, mortality, gender deformities, and health data is still very hard to obtain. Mayak is also involved in the reprocessing of spent nuclear fuel from Russian reactors across many countries and from Russia's nuclear submarines. A big part of Russia's sale pitch to others is the ability to reprocess their spent reactor fuels. The Mayak facility adds to the economy of Chelyabinsk. But this reprocessing also generates a great deal of radiological waste. The fate of the resulting radioactive materials is not known. The Fissile Material Storage facility at Mayak is a fortress with walls 23-feet thick and was paid for by Russia and the United States, the U.S. contributing more than $300 million. The U.S. is said to have provided equipment, expertise and training to this facility including increased accounting and security, but now Russia has blocked all access to the Ozyorsk plant. The federal budget for strategic nuclear weapons in both the U.S. and Russia especially since the end of the Cold War has steadily increased. Both General Groves and Beria knew compartmentalization was an important means for keeping secrets and suppressing bad news. An important way to neutralize the plutonium disasters has been to "naturalize" them. Thus, Hanford, Mayak, Chernobyl, Rocky Flats, Los Alamos and other nuclear facilities have been repurposed as wildlife preserves or national historic sites. [1,69,70,71]

The Russian state made it nearly impossible to claim any liabilities related to radiation exposure in the Mayak area. The Mayak accident released some 20 million curies of radiation from the September 29, 1957 explosion, but this represented only about 1/6 of the overall amount of long-living radionuclides released by the Mayak plant since 1949. That amount was estimated at 123 million curies.[4] Over four decades, Mayak and Hanford each discharged into the environment the equivalent of four Chernobyls. A dubious legacy continues to surround both plants today. In July 1945, the new President Truman revealed to Stalin news of the atomic bomb developed by the U.S. But Stalin knew of U.S. involvement in this area as early as 1942, thanks to espionage.[4,68,69,70,71]

7

ROCKY FLATS, COLORADO PLANT

Early Years

The Rocky Flats Nuclear Weapons Plant located about 16 miles upwind of the Denver Metropolitan area was a foundry that smelted plutonium, purified it, and shaped it into plutonium "triggers" for nuclear bombs and weapons. The plant also recycled fissionable material from outmoded bombs and weapons. It was the only plant early-on in the AEC/DOE system that produced triggers or "pits"- spherical plutonium explosives that initiated an atomic bomb's chain reaction. Construction of Rocky Flats was started in May 1951. The plant was opened in 1952 without the knowledge of Denver Metropolitan residents until much later. The plant employed thousands of workers. Admission that the plant was routinely handling plutonium was not announced until June 1957. A regional map of the plant is shown in Figure 7.1, and an immediate site map in Figure 7.2. Rocky Flats as part of the U.S. Weapons Complex is described in Figure 7.3. Rocky Flats was a cluster of shabby gray concrete buildings with a distinct government feel.[7,19]

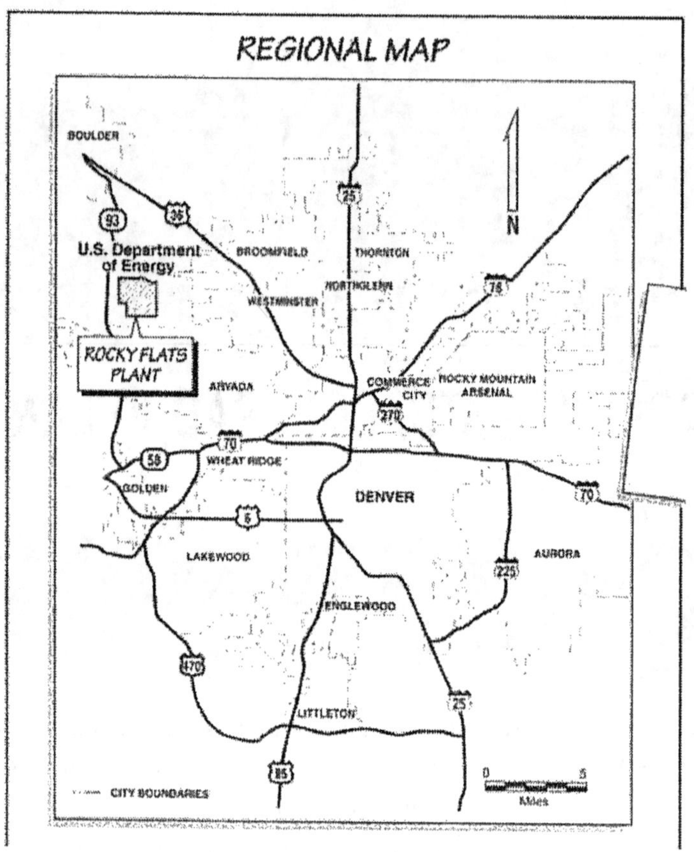

Figure 7.1 Rocky Flats Nuclear Facility Regional Map, 1953–1992

Credit: Wikimedia Commons and local.gov

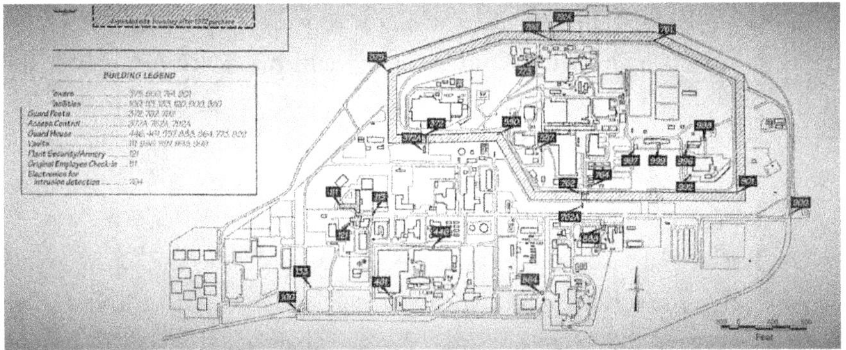

Figure 7.2 Rocky Flats Nuclear Facility Plant Security and Site Map

Credit: Wikimedia Commons and local.gov

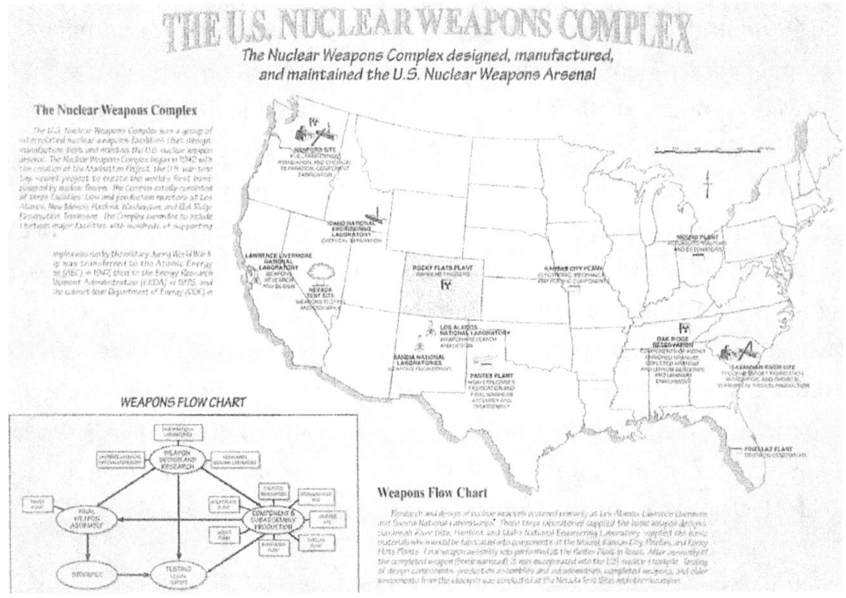

Figure 7.3 The U.S. Nuclear Weapons Complex and Rocky Flats Plant

Credit: Wikimedia Commons and local.gov

Rocky Flats had a sister plant which was Mayak near Chelyabinsk in Siberia, Russia which manufactured, refined and machined plutonium for nuclear weapons, just like at Rocky Flats. Deeply contaminated, Mayak became the site of one of the worst nuclear accidents in history- an explosion in 1957 that released 50 to 100 tons of high-level radioactive waste contaminating a huge territory in the eastern Urals.[19]

At Rocky Flats, there was complete secrecy regarding the job one was working in- and workers were told not to talk under threat of losing their jobs. The material they were working with had a half-life of 24,000 years and was extremely volatile. The workers handled plutonium using heavy leaded gloves inside of glove boxes. If a worker came in contact with plutonium, their skin was vigorously scrubbed with Clorox and brushes. Between 14.2 and 29 tons of plutonium was said to have been spread across the site, much of it in the 400 miles of ducts and pumps and in outdoor hot spots. Unfortunately, very poor construction records, drawings and specifications were kept.[72,73]

Many workers at Rocky Flats during the 1950s and 1960s became very sick, but minimum attempts were made to locate these people and their adverse health effects. The waste coming out of processing tanks were placed in plastic bottles which were said to become criticality objects. The 903 Pad containing thousands of deteriorating and leaking drums became a major source of contaminated groundwater moving west and north of the main plant. The Great Western Reservoir and Standley Lake which were used as drinking water supplies and for recreation by nearby communities, were reportedly found to contain plutonium in their bottom sediments. Between 40,000 and 70,000 plutonium "pits" or chips were manufactured by the Rocky Flats plant as triggers for atom bombs and similar weapons.[7,8,19,72,73,74]

Plutonium Pits

Plutonium pits were the core of the implosion nuclear weapon and the atomic bomb dropped on Nagasaki, Japan in August 1945. They are the heart of every nuclear weapon made in America. The principal component in these bombs is plutonium-239. Some weapons tested during the 1950s used pits made of uranium-235 alone, or in composite with plutonium. It was reported these triggers or pits contained enough breathable particles capable of killing every person on earth.[19] Figure 7.4 shows a plutonium pit inside a tungsten carbide container. The cladding of the pit was important. The older pits contained about 4 to 5 kilograms of plutonium-239 whereas the newer ones were about 3 kilograms. The diameter of a modern pit is only about 4 inches although the earlier ones were larger. Efficiency of the pits was enhanced by injection of deuterium and tritium into the cavity of the core immediately before implosion. There are many different kinds of pits based upon the weapon design. Beryllium cladding was common, but presented risk for workers because inhalation of the dust could cause berylliosis. By 1996, DOE had identified more than 36 cases of chronic berylliosis at the Rocky Flats plant alone. Finished pits from Rocky Flats were sent to the Pantex nuclear plant near Amarillo Texas for assembly. With the Palomares B-52 crash in 1966 and the 1968 Thule Air Base B-52 crash, accidental dispersal of plutonium became of concern to the U.S. military.[19,74,75]

Figure 7.4 Sphere of Plutonium Surrounded by Neutron Reflecting Blocks
Credit: Wikimedia Commons, http://www.lanl.gov/misc/copyright.htm

Casting and machining of plutonium was difficult not only because of its toxicity but because plutonium has many different metallic phases, inhibited by alloying it with other substances as beryllium, gold, gallium, etc. Both uranium and plutonium are very susceptible to corrosion. Self-irradiation can also cause deterioration. After many years, another plutonium radioisotope- plutonium-241, present in small amounts in weapon-grade plutonium, can beta decay to americium-241, a powerful gamma radiation agent, with increased danger to nuclear workers. Before 1989, decommissioned nuclear weapons were sent to the Rocky Flats plant where the plutonium was removed with acids, a process that created very large amounts of plutonium contaminated wastes. After the closure of Rocky Flats in 1989, DOE relocated beryllium operations to the Los Alamos National Laboratory (LANL) together with pit production in 1996. LANL pit production was limited to 20 units per year. The military desired to expand this capability, but Congress repeatedly declined funding. Some 15 tons of plutonium were found in strategic reserve weapons and in surplus weapons that were stored at Pantex. The pits at Pantex were enclosed in canisters in shielded bunkers until a better storage solution could be found.[11,19,74]

Operations at Rocky Flats

A glove box was where these plutonium triggers were made. The production line at Rocky Flats comprised a series of linked, stainless-steel glove boxes, up to 64-feet in length, in which plutonium was shaped by human hands inside protected gloves. The glove boxes were kept at a slight vacuum to preclude plutonium particles from escaping into the worker space and the environment. Workers who were clad in uniforms stood in front of the glove boxes and placed their arms into heavy, lead-lined gloves and peered through an acrylic window to mold and hammer the plutonium into its proper shape. An overhead conveyor system moved the plutonium from task to task along the line. It was difficult to man these stations without attendant risk since plutonium is highly combustible. In all cases, production had a higher priority over worker safety.[19]

Since the 1950s, Rocky Flats had been storing liquid low-level radioactive waste and sewage sludge in five shallow man-made ponds which were called "evaporation ponds" by Rocky Flats. In June 1979, a fluidized bed incinerator located in Building 776 for combustion of low-level transuranic waste made its first test runs. This incinerator apparently followed a 1958-built waste incinerator. "Transuranic" means artificial elements higher in the periodic table than plutonium as a byproduct when the latter is produced. The public was not informed of this operation until 1986, some 7 years later. There was serious objection to the incinerator together with a lawsuit brought by adjoining landowners for the contamination of 2,000 acres, which was settled for $9 million, but these records were sealed.[19]

Operating conditions in 1979 were rapidly deteriorating, and Jim Stone- a main engineer was selected to bring order to the plant. Stone's job was to prepare weekly reports identifying key problems and make recommendations as to how to fix them. Stone took issue with many things. There was far too much radioactive waste being stored at the plant. Drums and containers full of plutonium were piling up, and plutonium-contaminated liquid from the evaporation ponds was being sprayed onto fields of grass even in the winter when temperatures were as low as low as 7°F, resulting in runoff to ditches and waterways leaving the plant property. There was inadequate monitoring of water wells. Plutonium was being caught up in the ductwork of buildings especially Building 771, and the amount of MUF (material unaccounted for) was growing at an alarming rate. The company also was burning plutonium contaminated waste in an incinerator that was not properly licensed or filtered. The most worrisome problem however was "pondcrete" described as another desperate effort to deal with overflowing radioactive waste at the plant. The plan was to mix the toxic sludge from the evaporation ponds with concrete, then pour the toxic pudding into plastic lined cardboard boxes the size of small refrigerators and ship the boxes to the Nevada Test site for burial. In October 1982, Stone realized the cement would not harden and the plutonium would not stabilize.[19]

Major Fires

Over the course of 40 years there were more than 200 fires at Rocky Flats, the most dangerous- the September 11, 1957, and the May 11, 1969 fires. The 1957 fire originated inside a glovebox resulting in contamination of Building 771 with an estimated property loss of $818,600 in that-time dollars. Estimates of plutonium loss in the 1957 fire ranged between 500 grams up to 92 pounds. An explosion occurred during plutonium degreasing in

June 1964. A flash fire occurred in October 1965 spreading contamination through Buildings 76 and 77 (later renumbered as 776 and 777). On May 11, 1969, there was an epic fire disaster in the production Buildings 776/777 similar to the disastrous fire that occurred in Building 771 on September 11, 1957. A spark was caused by a plutonium chip inside a glove box and without notice turned into a vigorous flame.[7]

Water is used on a plutonium fire only as a last resort because water can cause plutonium to go "critical" with devastating effects. As plutonium burns usually by spontaneous combustion, it turns into a very fine dust which is intensely radioactive and remains in the air for long periods. Contamination on surface of a person's skin can be usually scrubbed off, but if a plutonium alpha particle is inhaled or ingested, it will emit a high, localized dose of radiation. Internal alpha emitters are more harmful per unit dose than gamma or X-ray radiation and will cause permanent damage especially to the lungs. A single microgram of plutonium, that is one millionth of a gram of plutonium, was considered by DOE to be a potentially lethal dose, and was determined to be the accepted "tolerance level" for workers.[7,19]

Carbon dioxide foam was first used on the 1969 fire but without success. The firefighters were feeling very warm and black smoke was coming out of the stack of Building 776. Water was subsequently used and there was an explosion. Most of the filters in the building were destroyed, and the lead cap was blown off the 152-foot smokestack of Building 771. Smoke poured out of the smokestack for 13 hours with flames 200-feet above the rim. Extensive firefighting with water ensued and there was grave danger of the roof over the buildings collapsing. Criticality was avoided. Nobody knows how much radioactive and toxic material escaped to the environment. AEC downplayed the risk of the 1969 fire to downwind residents, crops, etc. AEC did report that 41 workers endured substantial doses of radiation from the fire. There was little or no notice given to Denver residents, and everything was said to be safe. Contrary to Rocky Flats and DOE claims, the radioactive plume from the fire floated over the cities of Arvada, Golden and Wheat Ridge, and then passed to the north side of Denver and beyond. DOE described a massive decontamination program which lasted 2 years, with damages of $26.5 million. Another report said there was a $71 million loss.[7,19]

Following the 1969 fire, a nuclear chemist- Ed Martell made a request for information from Rocky Flats on the fire. Martell was former director of the Armed Forces Special Weapons Project studying the effects of detonation of the U.S. nuclear bomb tests in the 1950s and then working at the National Center for Atmospheric Research in Boulder, Colorado as head of a group called the Colorado Committee for Environmental Information (CCEI). His request was denied. The military further added that there would be no off-site testing. Martell's group then took soil samples from 2 to 4 miles east of the plant and tested for plutonium-239 and strontium-90. The CCEI completed its report in February 1970 and nearly all its soil samples showed plutonium originating from Rocky Flats. Plutonium in the top centimeter of soils was up to 400 times average background levels, and in some cases up to 1,500 times higher. Plutonium was present down to 5 inches in soil depth indicating a long history of leaks at the plant. High readings of plutonium were also found in waters of Walnut Creek which feeds into Great Western Reservoir, the water supply for the city of Broomfield. Martell learned from the Colorado Department of Health they had known of escaping plutonium from the site for some time especially from previous fires and the massive waste drum storage area.[19]

The AEC conducted its own study and found plutonium released by the plant was far below permissible levels, and the public was safe. Martell estimated at that time 200,000 to 300,000 people lived immediately downwind from the plant, and plutonium had been found as far as 40 miles from the facility. He was particularly concerned about the suburbs of Arvada, Westminster and Broomfield. It was also learned Rocky Flats had no emergency response plan to protect against a major plant disaster.[19] In May 1990, a fire caused damage of more than $250,000. Another in-plant fire occurred in May 1991. A significant grass fire covering 160 acres occurred in the buffer zone near Highway 128 in March 1994, and another in September 1996.[7]

In May 2000, a major fire erupted in the plutonium building 771. The criticality alarm was going off, and the noise was said to be deafening. The fire was in a glovebox, which was as big as a refrigerator. Workers were still in the room without protective gear. The fire was finally put out with the use of water, but one front-line firefighter was extremely "hot" with radiation received. He was scrubbed for hours and received tests for internal contamination which went on for weeks.[19]

Other Studies

In 1973, Al Hazle of the Colorado Department of Health (CDH) had returned from the Rulison Project- a 43 kiloton nuclear test project in western Colorado intended to extract natural gas from deep underground where he had been collecting water samples to test for tritium. He decided to take samples of water from one of the canals near Rocky Flats as background for his Rulison samples. To his surprise he found tritium in the runoff from Rocky Flats. But unless there was criticality or a nuclear explosion at Rocky Flats, there should have been no tritium anywhere near the plant. In 1974, Dr. Harvey Nichols at the University of Colorado and his research team collected snow and soil samples under an Energy Research and development Administration (ERDA) grant from an area some 3 miles east of the Rocky Flats Plant and found radioactive particles at all levels. Plutonium was also found in the top layer of snow suggesting new plutonium coming from ongoing emissions at Rocky Flats. When Nichols asked if plutonium was being emitted through the plant's stacks as part of its routine operations, the answer was yes, but the facility said there was no cause for concern. No further action was taken, and the government decided not to pursue further testing.[19]

Dow Chemical and the AEC did no further water sampling before saying Rocky Flats had no source that could possibly account for the tritium contamination. CDH tested the water again and showed that radioactive tritium released from Rocky Flats between April 1969 and September 1974 had entered Walnut Creek and flowed into Great Western Reservoir, the primary source of water for Broomfield, and the EPA confirmed these results. AEC was slow to acknowledge that a tritium leak had happened and said tritium levels were far below harmful conditions, but Ed Martell disagreed with this latter finding. Broomfield's reservoir water showed 23,000 picocuries of tritium per liter whereas AEC itself considered normal background to be 1,200 picocuries per liter. Martell claimed even the latter level was unsafe. A picocurie is one trillionth of a curie. An EPA report said that 50 to 100 curies of tritium eventually reached the Great Western Reservoir. Rocky Flats told residents the plutonium in the reservoir was harmless as long as it remained in the bottom sediments, and the same was true of Standley Lake. Rocky Flats eventually said their tritium was caused by scrap material shipped to them from the Lawrence Livermore Laboratory in California.[7,19]

In 1973, Dr. Carl Johnson, a professor at the University of Colorado became Director of the Jefferson County Health Department. In 1981, he published a study on high cancer rates in three "exposed" areas around Rocky Flats. Nobody believed to believe Dr. Johnson. Rocky Flats continued to insist that their buried waste posed no hazard. But tests confirmed that plutonium, americium and strontium-90, by-products of a nuclear explosion, existed in offsite areas. The presence of strontium strengthened the suspicion that a criticality did occur, perhaps during the 1957 fire. Around this same time, a newspaper article said another radioactive element was quietly in use at the plant- curium, which is 300 times more toxic than plutonium, but Jefferson County was not informed. The Jefferson County Commissioners came to Johnson asking for his approval on a new housing development just three miles from the plant, and Johnson said instead a study must be conducted to measure levels of radioactivity in breathable dust on the surface of the soil. Results were worse than expected. Plutonium in 1975 and 1976 was 44 times higher than what had been measured at the same locations by the CDH using the method of sampling whole soil, not surface dust. Johnson criticized the state's sampling methodology. The local planning board then vetoed the proposed housing development. Rocky Flats, local builders and others strongly opposed Johnson on this issue. The local population and the city of Broomfield were unhappy over the matter of contamination, and Johnson faced a growing storm.[19]

In early 1974, the Rocky Mountain News had the headline… "Cattle near Rocky Flats show high plutonium level as found by the EPA". Cattle just east of Rocky Flats had more plutonium in their lungs than cattle grazing on land at the Nevada Test Site. Also, plutonium, uranium, americium, tritium and strontium were found in measurable quantities in the cows' bodies. Officials at Rocky Flats challenged these results and one month later, the EPA said the conclusions were based on too little data. In the meanwhile, the resident population was growing ever closer to Rocky Flats.[19]

Rockwell International was selected in 1974 as the new overall contractor for Rocky Flats because the previous contractor- Dow Chemical no longer wanted the job. In 1974 after many public meetings, the Lamm-Wirth Task Force published a report which concluded not only that there were serious safety issues at Rocky Flats, but there was the potential for a catastrophic nuclear accident. They also said that Rocky Flats should be closed or relocated. It criticized the Price-Anderson Act which indemnified the nuclear industry against nuclear accidents and penalties even when gross corporate negligence was found. Ironically, the report emphasized that "strong consideration should be given to maintaining the economic integrity of the plant, its employees, and the surrounding communities". Some said the report was a masterpiece of compromising… "the health of local citizens with the competing interests of a government that wants to make bombs, developers that want to sell houses, and workers who need jobs."[19]

With funding from the NIH and growing opposition from nearly every side, Dr. Carl Johnson continued his studies downwind of the Rocky Flats plant in 1975-1976 finding plutonium in various environmental media. Johnson examined the relation between elevated cancer rates and exposure to plutonium for people living near Rocky Flats. His results were published in the Science magazine in 1976 and confirmed by international scientists. The government questioned his sampling methods, and the debate was lost in a flurry of negative publicity. Rocky Flats officials and the CDH in a scathing editorial attacked Johnson in the Denver Post. Years later in 1990, a reporter for the Denver Post uncovered the story that Rockwell paid a cash bonus for persuading the Denver Post to publish the editorial.[19]

By 1978, news of Dr. Johnson's research began to reach the public. He demonstrated a stark pattern of "excess incidence of all cancer in all categories" for both genders in areas exposed to Rocky Flats contamination. Johnson said in 1980 the incidence of cancer in Rocky Flats workers was alarming. Besides plutonium, other things showing up in the environment included cesium, curium, strontium and carbon tetrachloride. In January 1980, the EPA admitted to the press, it believed cancer deaths from Rocky Flats contamination might be occurring among Denver residents. In reports prepared by DOE itself based on air monitoring stations at Rocky Flats, higher levels of plutonium were found than at any other place in the western hemisphere. The question was how to keep someone like Dr. Johnson from releasing even more alarming reports to the press. In May 1981 the politics changed in Jefferson County, and Johnson was forced to retire from his job. In 1988, Dr. Johnson died from complications of surgery at the age of 58, never receiving any respect from state and federal governments.[19]

Protests

April 1978 was the beginning of the year of protests at Rocky Flats. As many as 6,000 persons showed up on April 28, 1978, to demand Rocky Flats be moved or shut down, and on May 7 there was a repeat performance. On April 28, 1979, protesters showed up again at the plant, and 286 were arrested. In October 1983, some 15,000 protesters attempted to completely encircle the plant, but did not quite succeed.[19]

FHA Notice of Proximity to Rocky Flats

In 1979, the FHA established a legal requirement that anyone buying a home within 10 miles of Rocky Flats with FHA insurance must be informed of plutonium contamination in the area. It said one should be aware that there exists within portions of Boulder and Jefferson County varying levels of plutonium contamination of the soil- but the EPA said these levels were safe. In 1982, HUD suspended the Rocky Flats Advisory Notice. The developer involved in getting the Notice lifted had invested in at least two housing enterprises near Rocky Flats. And the government continued to press the plant to make plutonium pits for at least another decade.[19]

Plutonium Storage and Pad 903 at Rocky Flats

Waste plutonium storage has been a serious problem at various DOE nuclear weapons facilities. An internal review by DOE in 1994 found significant hazards from leaky packaging, decaying buildings, piping and machinery at 35 government sites in more than 12 states. This study did not account for plutonium being stored inside warheads and pits, and Rocky Flats was found to still be housing 14.2 metric tons of plutonium at its site.[11]

A substantial part of the total problem was described as "mixed wastes." Daunting quantities of various mixed waste materials must be disposed of properly, simply because they have been in contact and contaminated with radioactive elements. In the long run, disposing of weapons-grade plutonium may be an easier task than getting rid of all the mixed wastes. When plutonium is part of a complex mass, one does not know what all the components are, and therefore, all of it must be treated and disposed of as if it were the most highly- radioactive component of the mix. Contaminated objects must be disposed of in their entirety. One example is the contaminated glove box which can be compacted and then put into barrels for disposal. Disposal can cost twice as much as a new $100,000 state-of-the-art glove box. Mixed wastes include entire buildings at badly contaminated sites such as Rocky

Flats and whereby everything in that structure – pipes, floor tiles, paint on the walls and tremendously more- are considered contaminated mixed wastes. By far, the largest portion of DOE's weapons-grade plutonium outside of Pantex, Texas in the late 1990s, was located at Rocky Flats. Placing these things in a nuclear dump was expensive, even if one had such a dump to throw things into, because the nation lacked the ability to provide such a safe and secure storage facility.[11]

Since 1954, The 903 Pad at Rocky Flats, which covered an area of 260,000 square feet, had stored thousands of rusted oil barrels and other containers- more than 5,000 of them, completely open to the elements, because there was no other place to put them. Each 30 to 50-gallon drum held waste oil, solvents contaminated with plutonium and uranium, and were heavily radioactive. Building 776 was where radioactive waste in drums overflowed. Spent nitric acid and plutonium-bearing wastes were present. The containers for these wastes were said to be designed not to last more than one year, but the containers were in use far beyond that time. Tanks and drums were sitting everywhere around Building 771 with spent nitric acid among many other wastes. Safety was not of high priority, as compared to maximum production and secrecy. These contaminants leaked into the soil and the groundwater that fed into Woman Creek and then Standley Lake and the Great Western Reservoir. The word got out that numerous rabbits on site were "hot" which management knew as far back to January 1962 as shown in a top-secret Rocky Flats memo. The rabbits when dissected for analysis showed high levels of alpha radiation, especially in their hind feet.[19,72,73]

It was reported that one microgram of plutonium, i.e., a millionth of a gram of plutonium, can produce a fatal cancer according to standards set by the AEC as early as 1945. The physicist- Fritjof Capra said plutonium should be contained and isolated for 500,000 years, and many scientists believe there is no safe level of exposure to plutonium. Martell and his group were surprised to learn that AEC and Dow Chemical knew about the leaking drums and the spread of contamination for at least a decade, but they kept this information secret. Over time Rocky Flats removed most of the barrels and covered a portion of the 903 Pad with asphalt. Some of the barrels were sent to a waste site in Idaho, and some were buried onsite possibly contributing to groundwater contamination. Even after the removal of the drums, the powerful winds in the Rocky Flats area were capable of suspending a large portion of the plutonium as dust which could be carried toward Denver. Winds in this region can reach 125 mph or greater. Figure 7.5 shows a Colorado Department of Health graphic of the path of airborne plutonium from Rocky Flats moving toward Denver correlated with the lifelong risk of cancer over the Denver Metropolitan area. Various scientists suggested that federal safety guidelines for low-level exposures to radiation be reduced by 90%, and AEC immediately challenged the findings.[19]

DOE Report of November 2017 on History of the Rocky Flats Nuclear Weapons Production Facility Close to Denver.

This report gave a long list of important events, but in several cases, it lacked specifics on site operations, accidents, and plutonium and radioactive waste releases in order to satisfactorily assess the cleanup program during the 1990s and the early 2000s. It also offered little detail on the Grand Jury Investigations of 1989-1992. A large-scale leakage was found coming from Pad 903 containing some 3,570 drums of lubricants and oils laced with plutonium, the drums having been stored from 1958 until their final removal in June 1968. The pad was covered with gravel after groundwater contamination was found, and the pad then coated with asphalt which probably was less than Best Management Practices.[7]

In April 1996, a moratorium was placed on the destruction of all records at Rocky Flats including those at the Denver Federal Center and Albuquerque. In May 1998, the EPA fined the DOE $45,000 for two violations of surface water standards in Walnut Creek caused by Rocky Flats. Then in August, the CDPHE and EPA fined Rocky Flats $489,000 for delays in draining radioactive and hazardous tanks. Estimated cleanup cost of Rocky Flats as of 2000 was expected to rise to $7.7 billion.[7]

LIFETIME CANCER RISK FOR THE LABORER, ALL EVENTS

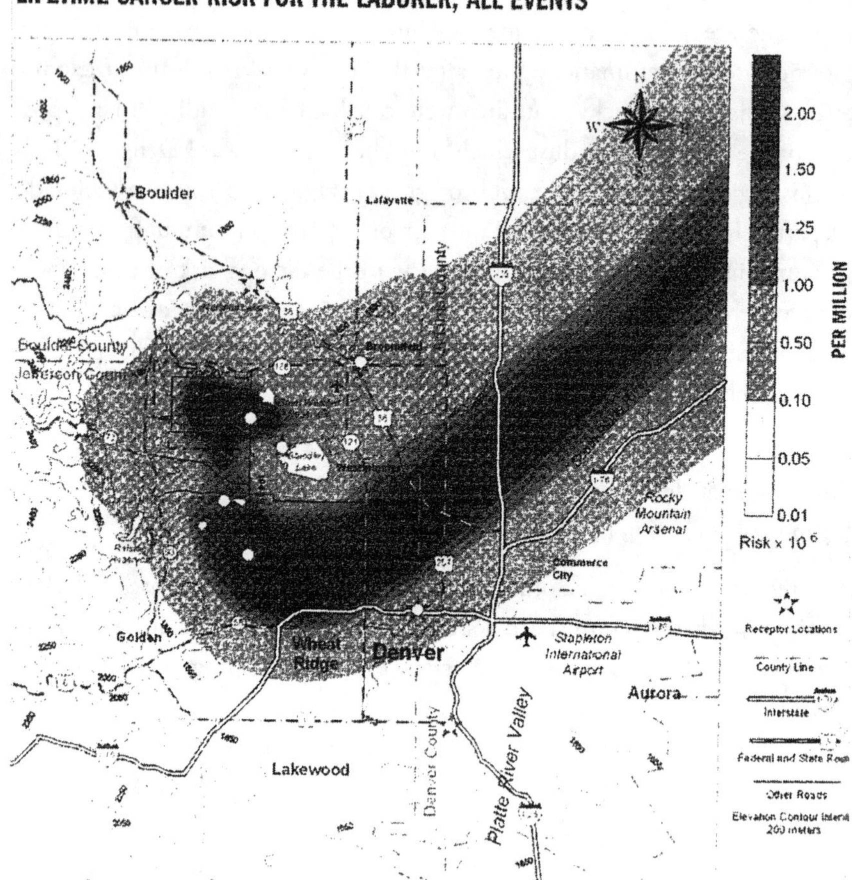

Graphic showing path of airborne plutonium from Rocky Flats and the risk of cancer to a laborer in the affected areas as a result of the Plant's operations from 1953-1989. *Courtesy of Colorado Department of Public Health and Environment, Summary of Findings, Historical Public Exposure Studies on Rocky Flats. (August, 1999)*

Figure 7.5

Pondcrete and Other Environmental Issues

Cleanup of the wastes in the five Rocky Flats evaporation ponds was started using concrete mixed with sludge and liquid to produce "pondcrete", despite Jim Stone's repeated memos that the process was not working and the material failed to solidify properly. The workers called this material the "Jelly Factory." Pondcrete and saltcrete wastes were exceeding available capacity and being re-processed and repackaged for storage and eventual offsite disposal. The result was 12,000 one-ton non-stabilized blocks which were allowed to stand out in the open unprotected. It was said the boxes were piled on top of each other like huge, soggy, sagging Lego blocks. In less than a year, the blocks were disintegrating, with their contents running downhill toward Walnut Creek and Woman Creek. By the end of 1986, some 2,000 blocks of pondcrete had been shipped and buried at the Nevada Test Site, but suddenly shipments were halted because the Nevada facility was found not to have a permit for receiving mixed waste. After RCRA and the Clean Air Act became law, these nuclear facilities had to break the law because production could not stop. Rockwell told Jim Stone to withhold his reports from the DOE. And then one day, Jim Stone was given one hour to leave the plant, and he would never again work in the nuclear weapons industry. In January 1990, pondcrete processing was further expanded. Eventually, 16,500 one-ton pondcrete blocks would be stacked on an asphalt pad just upwind of Arvada, Broomfield and the city of Westminster.[19]

Rocky Flats wastes were being shipped in various directions to the Waste Isolation Pilot Plant (WIPP) plant near Carlsbad, New Mexico; the Idaho National Laboratory west of Idaho Falls; the Nevada Test site; Oak Ridge; Envirocare of Utah; Hanford, Washington; Los Alamos; Erie Ohio; the Lowry Landfill; and other locations. In 1989, sanitary waste was going to Erie, Colorado because onsite landfill capacity was reached.[7,19,77,78,79]

Around 1989, a comprehensive DOE study of 160 contaminated sites at 16 nuclear weapons facilities across the country was released to the public, and Rocky Flats was declared to be the most dangerous site in the U.S.

principally due to hazardous waste in the groundwater and the large population directly downwind and downstream of the plant. Building 771 made the list of the 10 most contaminated buildings in America.[19] The railcar still sitting at Rocky Flats after its refusal to be accepted by Idaho was said to have 200 grams of plutonium in each of its 140 drums going nowhere. Besides groundwater contamination at Rocky Flats, there was plutonium contamination of soils and sediments, and the "solar evaporation ponds" further contributed to contamination of water supplies. An investigator for the U.S. Senate's Governmental Affairs Committee, which was responsible for monitoring nuclear weapons complexes said that groundwater and soils at Rocky Flats were so full of radioactive materials and toxic chemicals that Rocky Flats was likely to become a "national sacrifice zone," an area that will remain toxic for so long that no living creature could enter without endangering its health.[19]

Other DOE sites were also in trouble. At Hanford, Washington, liquid radioactive and toxic wastes dumped into trenches for four decades had contaminated large underground reservoirs used for drinking water and irrigation. Large amounts of radioactive materials had been released to the Columbia River and into the air, including some 740,000 curies of I-131, leading to higher incidence of thyroid cancer and other thyroid conditions. The Nevada Test site, the Pantex facility, and the Savannah River plant were said to be "deeply" contaminated.[19]

Idaho National Engineering Laboratory (INEL) Receives Large Amounts of Rocky Flats Waste Over Many Years

Information became available on radioactive wastes shipped from Rocky Flats to the INEL located 20 to 50 miles west of the city of Idaho Falls. INEL takes nuclear fuel and various waste products from nuclear plants and reprocesses these materials for recovery of uranium and other valuable commodities. INEL is the largest inland area of the federal nuclear weapons facilities at 570,000 acres. It was known earlier that the Governor of Idaho in October

1988 had refused to accept any more railcars of Rocky Flats waste and returned a shipment back to Rocky Flats where it stood on the tracks until early 1989 when clearance was received to accept it at the Waste Isolation Pilot Plant (WIPP) in New Mexico. Rocky Flats nuclear waste had been sent to Idaho over the period of 1954 to 1970, and buried in spacious unlined pits and trenches at INEL which overlaid a giant underground aquifer- the Snake River Plain Aquifer, a Lake Erie-sized aquifer serving farms and cities in the region. Contamination reached the aquifer according to a USGS report by means of injection wells, unlined percolation ponds from the above-cited pits, and accidental spills mainly during the Cold War era. Tritium accounted for much of the radioactivity in liquid entering the aquifer, but also included strontium-90, cesium-137, iodine-129, plutonium isotopes, uranium isotopes, neptunium-237, americium-241, and technethium-99. A 2020 USGS report said contamination in the aquifer has stayed the same or decreased, and was mainly below drinking water limits.[7,13,16,77,78,79]

The majority of transuranic or TRU wastes at the INEL came from the Rocky Flats facility, but also came from other sites. TRU wastes are materials contaminated with artificially-made radioactive elements such as solid sludge, clothing, tools, rags, soil, debris, etc. The name "trans" means beyond uranium, and may include plutonium, neptunium, and americium. INEL and the Hanford, Washington sites housed the greatest volume of TRU wastes in the United States. Rocky Flats wastes from 1954 to 1970 were packaged in storage drums and boxes. Originally there was also 65,000 cubic meters of TRU waste "retrievably"-stored above ground at INEL. In 1989, INEL became a Superfund site. DOE shipped nuclear waste to Idaho until a series of lawsuits were filed by the state against the federal government in the 1990s which led to a 1995 settlement agreement. Idaho did not wish to become a high-level nuclear waste repository.[77,78,79]

Agreements between the state and DOE were made in 1995 and 2008 for the cleanup and removal of wastes from the Idaho site. The buried waste was to be removed from a 97-acre landfill at the 890 square mile site which included the INEL. The targeted radioactive Rocky Flats wastes were said to include plutonium-contaminated filters, graphite molds, sludges containing solvents and oxidized uranium. Cleanup started at INEL in 2005. Both aboveground and subsurface buried wastes were to be removed to the WIPP, New Mexico facility under the Advanced Mixed Waste Treatment Project (AMWTP) and another program. The AMWTP is one of about 12 different cleanup efforts of nuclear waste finished or ongoing at this DOE site. The subsurface wastes were of first priority and were to be dug up or exhumated from 5.69 acres of the subsurface disposal areas. DOE said it removed about 13,500 cubic yards of material which is the equivalent of nearly 50,000 storage drums each containing 55 gallons. Delays have been experienced in the cleanup program due to accidents at INEL and at WIPP. The original cleanup completion date was December 2015, was said to be nearly finished, but is now extended to 2022 if not beyond .[77,78,79]

FBI–EPA "Sting" Raid of Rocky Flats on June 6, 1989

On June 6, 1989, 80 to 90 agents from the FBI and EPA arrived at the Rocky Flats plant to carry out a warrant to search for evidence of alleged criminal violations of RCRA and the CWA. They were onsite until June 23. On June 7, all production at Rocky Flats was shut down, and processing lines with plutonium were stopped in various stages. The evidence indicated that for over 30 years, spills, leaks, and waste disposal practices contaminated many sites around the plant. The most critical issue was groundwater pollution. Groundwater at Rocky Flats is quite shallow, only 25 feet below the surface. The sites where groundwater was most seriously contaminated were the 881 Hillside, the Mound Area, and the East Trenches. A disturbing discovery was that more than 2,640 pounds of plutonium was missing. Although Rocky Flats had always insisted that there had never been a criticality at the plant or an incident of uncontrolled fission, one memo reported an average of two nuclear criticality infractions per month. Surprising were the huge blocks of pondcrete which had been sitting in the open, the public not realizing these were illegal and not covered by permits.[19]

As a consequence, a Special Grand Jury was convened to review allegations and evidence secured during the raid of June 1989. It appeared the plant was frequently moving waste inside the plant from one location to another to avoid discovery. Also, plutonium pits were being moved from Rocky Flats to Los Alamos, Pantex Texas, Savannah River, and Lawrence Livermore in California. In November 1989, Rocky Flats officially ceased production, but certain processes continued.[7,19]

In 1992, process operations essentially ceased at Rocky Flats, its buildings were demolished and buried four feet below ground surface aka "buried infrastructure." Although cleanup was said to be complete on October 13, 2005, it was contended the local population and surrounding environment were seriously impacted. Worker exposures were poorly or not documented in good part because short cuts were taken and workers in heavy suits could not move fast enough to keep up with production goals. Although Rocky Flats today constitutes a wildlife area, many believe the site continues to be heavily contaminated having long-term adverse consequences, and the land is poisoned. The site remains a potentially serious environmental liability.[72,73]

After the raid, there was a class action lawsuit against Dow Chemical and Rockwell, involving some 12,000 property owners living within 5 miles of the plant, who claimed their properties had been contaminated by Rocky Flats. Wes McKinley- who we will meet later in this chapter said… "The State Health Department is very political … and never was on top of what was going on at Rocky Flats". The legal case was filed in 1990, and not until 2006 after many of the original claimants had died, the plaintiffs were awarded punitive damages of $110.8 million against Dow and $89.4 million against Rockwell with an additional $177 million in actual damages against each entity, for a total judgment of $554.2 million. The defendants said they would appeal. In the spring of 2010, four years later, the appeal went before a three-judge panel in the 10th Circuit Court of Appeals in Denver. The award was now $926 million including damages and accrued interest. This case had been in litigation for nearly 20 years. On September 2, 2010, the original decision was overturned predicated on insufficient risk and not being actual damage. Despite contradictory data, the DOE and the CDPHE continued to maintain that Rocky Flats was completely safe. As time passes, most people seem to have forgotten about Rocky Flats.[8,19]

The Special Grand Jury of August 1989 to March 1992

A summary of the Grand Jury actions is given first, then specifics are provided based on a 2004 book written by the Foreman of that Grand Jury[8] and a 2013 reference.[19] This information is available despite the fact evidence and findings may still remain sealed from the public by the Federal Judge presiding over the Grand Jury.[8] Members of the Grand Jury were silenced in their findings and recommendations, threatened, told they would be arrested if they presented their story, and that the government would not move forward on the case. Immunity was given to DOE officials and Rockwell executives after a multi-million-dollar fine was agreed to. Rockwell after offering great resistance settled, but no individuals were ever indicted. Members of the Grand Jury were highly frustrated. It was said the Rocky Flats cleanup could take up to 100 years.[72,73.]

The Special Grand Jury was in session generally for one week each month at least in the early stages. After the initial fact-finding phase went nowhere and was over in January 1990, members of the Grand Jury were surveilled by the FBI for speaking out, but they were eventually exonerated. The Jury Foreman was Wes McKinley who was unhappy over the situation. The Grand Jury took testimony from many witnesses and spent considerable time in deliberations. During testimony, the witnesses before the Grand Jury said they had been threatened. Many Rocky Flats workers were said to have been contaminated, and there were several cases of radiation sickness. Onsite drums of hazardous material containing oil and plutonium were rapidly deteriorating. Offsite contamination was concentrating in adjoining wildlife. There were frequent fires and accidents. During the major fire of 1969, more than one ton of plutonium could not be accounted for. Soils southeast of the plant were found to contain radioactivity more than 400 times ambient levels, and some results were 1,500 times higher. Thousands of drums were seen in the fields corroding and deteriorating.[75]

Rocky Flats had burned radioactive waste in an incinerator for years, and radioactive wastes had been sprayed onto surrounding fields. Supposedly the plant could not operate without proper permits but did so anyway. Following the Grand Jury trial, all evidence was sealed and there it stayed. Building 881 was a place where plutonium was accumulating in the air ducts and walls. Some 62 pounds of plutonium were found in the ventilation works. In the FBI-EPA raid of June 1989, 969 boxes of evidence were secured. Rocky Flats was trying to mix wastes with concrete, but it turned to "mush." The plant was spraying waste onto adjoining fields even in the wintertime. Waste was getting into the groundwater and surface streams being used by municipal water supplies downstream. Rockwell lobbied hard, and threatened to sue the federal government. The Grand Jury was planning a number of indictments against DOE and Rockwell individuals, but in December 1990, environmental data collection stopped, and the Grand Jury was told to go home. The last meeting of the Grand Jury was held in March 1992. A plea bargain was arranged with Rockwell who paid a considerable penalty, but no individual indictments were secured. Rockwell said there was no radiological danger outside the plant.[8,75]

The Grand Jury report was completed but suppressed-- still some of it got out to the public. It was said that a contractor-DOE culture prevailed, and there was no accountability in government. Years later, groundwater contamination had spread to adjoining farms, yet homes kept moving ever closer to the plant. And the initial plutonium health and safety levels in the plant were said to have been set 500 times higher than they should have been.[75]

The Federal Judge went against the claims of the Grand Jury members and sealed all the collected evidence which also said how much plutonium had been lost at Rocky Flats. The Colorado Department of Public Health said… "We didn't need to know", and that DOE was responsible. The DOE culture did not change, and the contractors continued to manage the situation. Nobody seems to know what actually was buried at Rocky Flats. The public knows much more about Rocky Flats because of the Grand Jury, but this is not true of other nuclear weapons manufacturing facilities. DOJ excused the actions of the Grand Jury but said further action would be taken if there was more talk.[74]

The most significant issue of the Special Grand Jury was probably the plutonium incinerator in Building 771 at Rocky Flats. In October 1988, the plant was ordered to shut down its plutonium incinerator because of a previous serious accident. About six months prior to the FBI/EPA raid of June 6, 1989, unbeknownst to Rockwell, the EPA out of Las Vegas conducted nighttime aerial reconnaissance using infrared imaging of the plutonium incinerator smokestack on December 9th, 10th and 15th, 1988 giving evidence that the incinerator was indeed operational. The FBI suspected both Rockwell and DOE officials of breaking environmental laws resulting in the June 6, 1989 "sting" raid. This was the first time the FBI had ever served a search warrant on the U.S. government. The raid was led by Jon Lipsky of the FBI. It seemed there was severe concern about Rocky Flats, even the federal government admitting in a 1988 study that groundwater contamination at the site was the worst in the entire national weapons complex. Jim Stone was a major whistleblower, an engineer who previously worked in utilities and ventilation and who confided in Jon Lipsky. Other important persons were Jacque Brever and Karen Pitts who were chemical operators in Building 771; Ron Avery, a Foreman in Building 771 who had left Rocky Flats two weeks before the raid for Florida; and Wes McKinley, the Grand Jury Foreman.[8]

On the second day of the raid, Jon Lipsky learned that the U.S. Attorney General Thornburgh gave a press release stating the objectives and strategy of the raid which was said to give Rocky Flats a significant advantage over the FBI-EPA. Rockwell reportedly intimidated workers into not talking to the FBI and told any whistleblowers they would be dealt with harshly. Lipsky was worried Rocky Flats had been given advance notice of the raid. One woman said the plant was playing a shell game with the personnel moving wastes and attendant paperwork from one location to another just ahead of the agents entering that location. Brever and Pitts were interviewed by the FBI on June

16. Shortly after, the FBI accidentally or otherwise divulged the names of the informants they had. Brever and Pitts both worked in the incinerator area in December 1988 but had not been told of the incinerator shutdown order.[8]

On June 13, 1989, all personnel in Building 771 were told by Rockwell... that everybody knew the incinerator did not run between October 7, 1988 and February 25, 1989, which Brever and Pitts knew was untrue. Shortly after June 17, several cars appeared at Brever's house and rocks were thrown at her window by Rocky Flats workers including Brever's father. Around July, the harassment had escalated to the point where Brever and Pitts were afraid for their lives although they continued to work at the plant. On September 14, Brever, while on the job, was apparently deliberately contaminated by her fellow workers with what may have been a lethal dose of radioactivity. Lipsky, when informed, was furious over these incidents and with the two FBI agents who had identified Brever as an informant. Lipsky was finally able to transfer Brever and Pitts to the "cold side" of the plant before their testimonies were scheduled to be given to the Grand Jury.[8]

Brever was the first of many witnesses who gave testimony to the Grand Jury. The Foreman Wes McKinley was keeping a personal extensive journal on all Grand Jury deliberations. Lipsky was told by federal prosecutor Fimberg that the Grand Jury had not believed the testimonies of Brever and Pitts, and they came across as troublemakers. This statement was in direct contrast to notes in McKinley's journal. An early witness was Al Divers, the EPA infrared expert from Las Vegas who said the evidence was clear Rockwell was definitely operating the plutonium incinerator on the three flyover dates in December 1988. Rockwell with use of the newspapers and DOJ insisted that this midnight burning never happened. Fimberg later told Lipsky that Divers did not fare well in questions asked by the Grand Jury. On November 30, an article in the Denver Post said Governor Roy Romer and U.S. Representative David Skaggs believed no midnight burning had taken place. It was obvious there was a leak of secret Grand Jury information and it was believed this leak had come from the Justice Department. Despite the article, Brever and Pitts continued to provide information on unwarranted activity and safety problems at the same time being continuously harassed by their co-workers. The Foreman of the Grand Jury, Wes McKinley, was outraged- all Grand Jury proceedings were confidential and not to be determined by politicians, the FBI or anyone else.[8]

On March 26, 1992, the Grand Jury members and everyone else found out from the press that DOJ had ended the Grand Jury investigation, deciding to settle the case instead of going to trial. The government prosecutors announced their plea bargain, saying the $18.5 million fine against Rockwell was a new record, and insisted there had been no real offsite harm caused by Rocky Flats. U.S. Attorney Mike Norton said… "I know of no evidence of any physiological or environmental harm at *all* from the operations of the facility." It was thought Rocky Flats and Rockwell had made a deal with U.S. Justice. Brever then resigned from Rocky Flats and went into hiding. Justice said whatever environmental impacts there were appeared to be substantially limited to inside the plant boundaries. Jon Lipsky reflected to the time of the search warrant affidavit which gave detail on a number of offsite impacts from Rocky Flats. Lipsky himself had been left out of the negotiations and the plea bargain from the beginning. Then Lipsky found out the Grand Jury had written its report and wanted to speak publicly about it, but it was sealed by the federal Judge. The Grand Jury members were under orders never to talk about the investigation, or face prison for contempt of court.[8]

After the government's plea bargain, Wes McKinley, the Foreman of the now defunct Grand Jury, sent a copy of the Grand Jury's final report to the federal Judge in hopes of unraveling the situation. Lipsky wondered what the Grand Jury could do, and he already had seen a copy of the report which someone had secured from the Grand Jury vault. McKinley knew from the newspapers that the Judge had sealed the Grand Jury report from the public. Then the federal Judge took a public swipe at the Grand Jury saying… "It is with great regret that the Court notes that the Grand Jury, having the opportunity to inform the public of the facts of Rocky Flats, failed in its duty." A few weeks later, someone leaked the Grand Jury report to a local newspaper- the Westword. Not all the report was printed, but the public now knew the Grand Jury had written indictments of individual DOE and Rockwell officials, and the government prosecutors had obstructed their work. Federal Judge Finesilver's reaction was prompt. He requested the Justice Department and the FBI to investigate the 23 members of the Grand Jury. This time it was not the polluters under investigation, but the Grand Jury.[8]

In November 1992, McKinley and his attorney announced to the public they would be sending a letter to President-elect Clinton asking that the Department of Justice be investigated. The letter went out signed by eleven other jurors and sent to nationwide newspapers and media. McKinley read his letter in a public forum but could not answer crucial questions because FBI agents were in the audience to arrest him if he did so. Congresswoman Patricia Schroeder supported the Grand Jury going forward. The Grand Jury members and their attorney sought immunity from Congress, but this was shot down by the defense contractors and Congressman David Skaggs from Boulder who said these procedures were unconstitutional. The stage is now pushed forward five years to 1997 during which time Wes McKinley had secured an attorney- Caron Balkany. McKinley continued to believe the public did not know how bad Rocky Flats really was, and the government was now ready to turn this site into a wildlife refuge with public recreation.[8]

McKinley and Balkany started with a Congressional Investigation made of Rocky Flats in 1992 and a resulting Wolpe Report, which had two major recommendations. The first of these was that Congress should investigate whether Justice's handling of the Rocky Flats case had been… based in part on a desire to protect the current administration from evidence that it had not changed DOE's culture, which had encouraged continuing environmental violations. The second recommendation was whether Congress wished to hear from the Rocky Flats Grand Jury. The Justice Department simply said the Wolpe Report was wrong, and Congress had no right to interview FBI agents or DOJ prosecutors. Congress got tired of stalling tactics, and subpoenas were issued to Norton, Fimberg and Lipsky. Lipsky testified there were indications that something had gone very wrong with the Grand Jury investigation of Rocky Flats, and the plutonium incinerator was definitely operating illegally. Lipsky said he also had been told by his superiors not to look for further evidence because the case was going to be settled. Concurrently, GAO found cases of possible unearned bonuses to Rockwell, and DOE had found a total of 230 environmental and safety violations. The Wolpe Report published in January 1993 found the Justice Department had bargained away the truth about what really happened at Rocky Flats. This report was quickly attacked by Justice as unfair and inaccurate, and the FBI then transferred Jon Lipsky out of Denver for a less desirable assignment.[8]

The price of cleaning up the Rockwell pondcrete/saltcrete disaster alone was estimated at over $100 million, but Rockwell did not lose a penny for doing a poor job on this, and actually made a profit. The spray fields were supposed to receive only effluent from the sewage treatment plant, but instead were receiving radioactive and hazardous wastes. There was an excess of contaminated liquid waste which the fields could not accept, so it ran into Walnut Creek and Woman Creek then into drinking water reservoirs for Broomfield and Westminster. Both Rockwell and DOE were aware of this situation. Rocky Flats did not have permits for their waste dumps but proceeded anyway.[19]

DOJ was not interested in letting the Grand Jury do their job. Fimberg was angry when told the Grand Jury was ready to issue its report with indictments against individuals, and Fimberg said this would never happen. On December 12, 1991, Norton told the Grand Jury there would be no report, no indictments, and it would be inappropriate for the Grand Jury to meet again. It was learned later that it was Justice and not the Federal Judge who vetoed issuance of the Grand Jury Report. On March 24, 1992, Justice "forced" the Grand Jury to write a report version accommodating DOJ rather than the original Grand Jury version. After a series of votes, McKinley- the Foreman of the Jury handed the Judge a version marked "A True Bill" which contained the original indictments against individuals. He also handed the Judge a version given to the Grand Jury by U.S. Attorney Norton which the Grand Jury had signed as "Not A true Bill." The Judge then dismissed the Grand Jury.[8]

A Citizens Investigation was formed led by Wes McKinley and Caron Balkany. The nuclear weapons manufacturing industry was accustomed to there being no independent oversight other than by DOE. There was also the Unitary Executive Policy in the White House which meant the EPA was not allowed to take DOE to Court no matter how it performed. All the EPA could do was to negotiate. It was found in 2001 that Rocky Flats had received permit authority to incinerate over 22,000 pounds of plutonium contaminated trash yearly. Then they received permission to incinerate up to 336,000 pounds yearly. Since 1970, they had incinerated more than 345 tons of plutonium-contaminated trash, which did not include the ten previous years of incineration.[8] (The question may be asked what percent of plutonium in the trash was removed by burning and capture, and what percent was discharged to the atmosphere?)

The Rocky Flats case begs the question… When does a contractor have enough influence to convince the U.S. government to change the picture to the American people- and to Congress? Broomfield was complaining about high nitrate levels in the streams coming from Rocky Flats that feed their water supplies. Broomfield then discontinued taking their drinking water from streams coming from Rocky Flats.[8,19]

Caron Balkany interviewed Al Divers, the EPA aerial infrared expert who flew over the Rocky Flats plutonium incinerator in December 1988, and Balkany secured the graphs and annotations that had been withheld from Balkany under FOIA request by the Justice Department for so long. Divers said when he gave testimony to the Grand Jury, he had never changed his position on the Rockwell operating incinerator. Divers said the heat signatures he took of the incinerator smokestack in December 1988 when Rockwell said the incinerator was not operating, were exactly the same as when the plutonium incinerator was fully operational later in February 1989. He also guaranteed in December 1988 the incinerator stacks were running hot, the incinerator was operating, and the exhaust was much hotter than room temperature. But the Justice Department said in their public and congressional statements that Divers' change of position was the main reason they dropped the midnight plutonium burning charge. Divers seemed perplexed as to why Justice was claiming he had changed his story, and he had no idea of the bombshell he dropped on the Citizens' Investigation.[8]

Caron Balkany also interviewed Ron Avery, a Foreman in Building 771, who left Rocky Flats two weeks before the raid for Florida. Avery related how dirty the filters were following the scrubber system on the air exhaust before being discharged to the smokestack and the atmosphere. Avery said the filters were always black and were frequently wet which represented a dangerous situation for downstream populations. In 1997, but even before, the U.S. government finally admitted that the radiation exposure of its workers at (various) nuclear weapons plants had not been properly monitored and its data was suspect. Ron Avery was told when he was working at Rocky Flats, the permissible legal dose for workers at Rocky Flats and other nuclear plants in 1987 was 5 rems per year. In 1990, the International Commission on Radiological Protection recommended lowering this permissible dosage to 2 rems per year because the danger at low levels of radiation exposure was three to four times greater than previously thought. But DOE never reduced the permissible dose for its workers.[8]

Ron Avery said on that fateful weekend in December 1988 that he, Brever and Pitts ran the plutonium incinerator the entire weekend on a special run because… "they wanted me to get rid of a lot of high gram count barrels of waste." This run started on Friday. "We can only do about 10 to 15 barrels of that stuff a shift. It took us through Saturday night to do what they wanted." This was conclusive proof of secret midnight plutonium burning and the Justice Department had been less than honest about why they had dropped the charges on the incinerator burning at that time- as did Rockwell.[8]

Rocky Flats Cleanup

The National Academy of Sciences issued a report in August 2000 warning of continuing problems at nuclear weapons manufacturing sites. "At many sites, radiological and non-radiological hazardous wastes will remain, posing risks to humans and the environment for tens or even hundreds of thousands of years. Complete elimination of unacceptable risks to humans and the environment will not be achieved, now or in the foreseeable future." Around 2000, the federal government was planning a modern plutonium pit manufacturing facility, new nuclear weapons were under design, and the Nevada Test Site was again being made ready for testing. There was talk of abrogating the Comprehensive Test Ban Treaty, and environmental and health considerations were once more being pushed back into second place, if that, and secrecy was back. DOE argued that making new plutonium pits was necessary because the pits might get old and not explode destructively enough. The prime motivation however for a new plutonium facility was to re-create a capability to mass- manufacture entirely new nuclear weapons that required pits of a new design.[7,8,9]

People who lived near Rocky Flats who led "active" outdoor styles were said to have the highest level of exposure to airborne plutonium. Walnut Creek and Woman Creek continued to flow offsite from Rocky Flats and eventually into reservoirs that used to provide drinking water for local cities. Because of contamination, the city of Broomfield built a new drinking water supply, and the Great Western Reservoir was no longer being used for city drinking water. Standley Lake, which may have received plutonium, continued to provide drinking water for local communities and to be used for recreation. Despite ongoing requests for further health testing, the DOE and CDPHE said low levels of radioactivity did not warrant an epidemiological study of residents around Rocky Flats.[19]

The "reported cleanup" of the Rocky Flats Nuclear Weapons Plant was said to have been completed in February 2006, and in June 2007, DOE transferred management of the Rocky Flats National Wildlife Refuge across to the Department of Interior.[7,9] The cleanup was described in an Accelerated Site Action Plan that would include a 130-acre cement cap over the central high-security, highly contaminated process areas. By 2010, most of the major buildings would be demolished and plutonium would be sealed behind thick layers of concrete. The Infinity Rooms would also come under that cement cap. The Infinity Rooms were called that because the radioactivity was so high it could not be measured by normal equipment, and no one could access them without extraordinary safety measures. Infinity Rooms were scattered through three separate buildings. One of these rooms in 1968 was forever welded shut. In Building 771, there were 17 Infinity Rooms. Also, Pac Rooms stored contaminated equipment from around the site, into which workers had to go through four separate airlocks to enter these rooms.[19]

When money for Rocky Flats cleanup seemed limited, the DOE, EPA and CDPHE changed their minds and made a deal with Congress to impose monetary limits on the cleanup. A deadline of December 2006 was set, and cleanup and closure costs were not to exceed $7 billion. Even the most cynical observers were surprised when the interim Radioactive Soil Action levels (RSALs) were announced at 651 picocuries per gram of soil, significantly higher than at any other site in the United States. If contamination levels were found below RSAL, no cleanup action would be taken, and there would be no cleanup of soils. This level is compared to other plutonium contaminated sites. The bomb test sites at Eniwetok Atoll and Johnson Atoll in the Pacific are much lower at 40 and 14 picocuries per gram of soil, respectively. The Lawrence Livermore National Lab in California is set at 10, and Fort Dix in New Jersey is set at 8. A part of the Hanford, Washington site, perhaps one of the most contaminated in the nation, is set at 34 picocuries per gram, and even the Nevada Test site is lower at 200 picocuries per gram of soil.[19]

Citizen and watchdog groups were alarmed, and a 15-month study was conducted by a well-known group of scientists, who recommended a level of no more than 35 picocuries of plutonium per gram of soil. This report was sent to DOE who never formally responded. Then the DOE, EPA and

CDPHE came up with a compromise. The top 3 feet of soil would be cleaned to 50 picocuries per gram. But the soil 3 to 6 feet below the surface would be cleaned only to a level of 1,000 to 7,000 picocuries per gram. Nothing below 6 feet would be cleaned up, even though considerable contamination remained below that level. In 2000, Congress proposed turning Rocky Flats into a wildlife refuge, and this Act passed in 2001. Of the $7 billion slated for cleanup, much of that would go for site security and relocation of weapons-grade materials. Only 7% would go for soil and water cleanup. Government officials claimed that background levels for plutonium due to fallout from atmospheric testing of nuclear weapons from 1945 to 1963, made the whole argument moot. Yet, the average background level for plutonium along the Front Range of the Rockies resulting from global fallout is reportedly only 0.04 picocuries per gram of soil. The government has no intention of permanently "closing off" its land. Their view is that turning former nuclear sites into nature preserves is "thrifty environmentalism. It would otherwise cost a fortune to clean up the site so that people could live there… but making it safe for wildlife-dependent public use is more affordable."[19]

Human health studies have been conducted around Rocky Flats and other DOE nuclear facilities which have shown alarming results. In 2000, the Energy Secretary acknowledged workplace exposure had harmed the health of workers in the U.S. nuclear weapons industry, and the Energy Occupational Illness Compensation Act was passed shifting the burden of proof of exposure to the government rather than the worker. During the production years from 1952 to 1989, more than 16,000 people worked at Rocky Flats and thousands more during the post-production years. Some 6,000 applied for compensation and hundreds have received compensation. Yet as of 2011, two out of three who have sought compensation have had their claimed denied. Cancer was common among Rocky Flats workers but even in the most "proven" cases, claims were denied. The cleanup itself generates waste, and for every pound of detoxified equipment, another 1.6 pounds of waste is generated because cleanup equipment and materials also become contaminated.[19]

Based on compromised cleanup standards, the Rocky Flats cleanup was declared "successful" in 2005, and the EPA in June 2007 certified the site. While the cleanup was said "complete", 2,600 pounds of plutonium were still missing. The problems at Rocky Flats are shared by former nuclear weapons sites not only in the U.S. but around the world. Fernald, Ohio, a former uranium processing facility, "released millions of pounds of uranium dust into the air and contaminated surrounding areas with radioactivity". Today that site is "permanently closed", and nobody can ever safely live there. As of 2013, there were some 25,000 plutonium pits in the U.S. stockpile: some 10,000 are in nuclear warheads; 5,000 in strategic reserve; and more than 10,000 surplus pits are stored at the Pantex plant near Amarillo, Texas. The DOE says aging plutonium pits may be unreliable and new pit production is necessary.[19]

In May 2017, the last of four meetings for the public to share plans for the opening of the Rocky Flats National Wildlife Refuge was held, and activists were present. One person said the meeting was just a dog and pony show put on by the government agencies. Another said… "safe enough is never safe enough" and she said… "why are we even entertaining the notion of opening up what was one of the dirtiest sites for radiation for recreation use?" The cleanup was a decade long, with a cost of $7 billion at Rocky Flats, and the federal government considered it adequate.[80]

DOE said they maintain covers over two old landfills, maintain three groundwater testing systems, and watch over four water (wastewater) collection systems. Many critics in attendance at the above meetings said the site was not suitable for recreation which was reinforced by the 1,200-acre Central Operable Unit in the middle of the complex still being an active Superfund site and off-limits to the public for safety reasons. Standley Lake continues to serve as city water supply. Jon Lipsky who led the FBI-EPA sting raid on Rocky Flats in 1989 that ultimately contributed to the Rocky Flats shutdown, said… "I can't believe anyone would want to build a plutonium playground on Rocky Flats." There have been contradictory health studies regarding human health results in 1988-1989. Surrounding communities do not fully support the concept of a "refuge" and two school districts have banned classroom field trips to the refuge.[80]

The author of "Full Body Burden" was of the opinion … that it is not possible to make nuclear weapons without creating tons of nuclear waste. And it is not possible to make nuclear weapons without sacrificing the health of neighboring citizens and the workers. There is no safe way to make the bombs, there is no safe way to dispose of the wastes, to transport the stuff, or store it. And the worst part of all is that we don't need any more nuclear weapons. We have enough to blow up the planet 20 times over. We don't need more bombs, but there is so much money to be made. Who is watching out for public health and safety all the time?[19] Since 1993, former Superfund sites have been returned to communities as soccer fields, golf courses, and wildlife refuges, thereby promoting human access to the sites. The EPA is working with private organizations like the U.S. Soccer Foundation, the National Football League, and major league baseball to obtain their support for turning Superfund sites into recreational areas. Jon Lipsky and his wife were whistleblowers who were retaliated against for their stance about what happened at Rocky Flats.[8]

8

OAK RIDGE, TENNESSEE PLANT

The Oak Ridge National Laboratory (ORNL) is located in eastern Tennessee on the Clinch River about 15 miles from Knoxville. It was established in 1942 by the U. S. Corps of Engineers as a vital element of the Manhattan Project, and originally known as the Clinton Laboratory.[77] During WW2, it was managed by the University of Chicago's Metallurgical Laboratory. The federal government secured 59,000 acres (92 square miles) of land in 1942. Some families were given only two weeks notice to vacate their farms that had been their homes for generations. Others settled there after previous evictions to make way for the Great Smokey Mountains National Park in the 1920s and the Norris Dam in the 1930s.[3] Over 400 buildings were constructed at Oak Ridge which formed one part of a vast production network. The best-known site was Los Alamos where physicists and mathematicians designed the atomic bomb and assembled their final components. Hanford, Washington and Oak Ridge, Tennessee were less well known, but were the workhorse production facilities of the Manhattan Project.[3,13,81]

X-10 Reactor at Oak Ridge

The X-10 Graphite Reactor was one of the first structures built at ORNL to show that plutonium could be created from enriched uranium and was the world's second self-sustaining nuclear reactor after Fermi's previous experiment at the University of Chicago- the Chicago Pile-1. The initial uranium ores came from the Belgian Congo and Canada. It was the first continuous operating reactor. Figure 8.1 shows the X-10 Reactor in operation. In 1946, the first medical isotopes were produced in the X-10 Reactor, these products then shipped across the United States. The X-10 Reactor was air cooled and served as a pilot plant for the much larger production facilities at Hanford, Washington. The reactor was 24 feet long on each side and weighed about 1,500 tons surrounded by 7-feet of high-density concrete serving as a radiation shield. It was supported by a chemical separation plant. The X-10 plant also contained the Clinton Laboratories. Figure 8.2 shows the X-10 Reactor later under demolition and cleanup.[81,82]

The biggest problem encountered with the X-10 Reactor was the corrosion and clogging of the uranium feed slugs entering the reactor channels resulting in fission products being lost from the system. After many attempts, a successful aluminum coating process was developed for incoming slugs. The reactor went critical in November 1943 with 30 tons of uranium, and power was eventually raised to 4,000 kw in June 1944 producing the first few ounces of plutonium out of the separation plant. The X-10 reactor produced plutonium until January 1945 when it was largely converted to research and superseded by the Hanford reactors. Attention was then directed to producing medical isotopes. Figures 8.3 and 8.4 respectively show plant site 101 and the Oak Ridge facility around 2021.[3]

Figure 8.1 Oak Ridge X–10 Graphite Reactor, Operational November 1943

Source: USDOE and National Park Service

Figure 8.2 Oak Ridge X-10 Reactor under Demolition, May 2021
Source: USDOE EM

Figure 8.3 Oak Ridge Site 101, 2021
Source: USDOE EM

Figure 8.4 Oak Ridge Facility at Present
Source: USDOE EM

Three Separation Plants

One of the earliest plants built at ORNL was the Thermal Diffusion separation plant or S-50. Plans called for building 2,142 48-foot-tall diffusion columns arranged in 21 racks. Inside each column were 3 concentric tubes. The inner tube made of nickel contained steam at 100 psi. with a temperature of 545°F flowing downward. The outer tube or pipe made of iron contained water at a temperature of 155°F flowing upward. The middle tube made of copper contained uranium hexafluoride gas. Partial operation of the S-50 separation plant began in September and by October 1944 it produced just 10.5 pounds of 0.85% U-235. After many leaks in the cooling water and various shutdowns, by March 1945, all 21 production racks were operating. Initially the output of S-50 was fed into the Y-12 separation plant. In March, there were three enrichment processes being run in series. S-50, the first stage enriched U-235 to about 0.8%. The K-25 gaseous diffusion plant was the second stage enriching U-235 to about 23%. The third stage, the Y-12 electromagnetic plant raised the U-235 to about 85%, suitable for nuclear weapons. The U-235 used in the Little Boy atomic bomb dropped on Hiroshima, Japan was about 85% enriched.[3]

Operations and Worker Health and Safety Concerns

Uranium production at ORNL was always considered by the Army to be on a "permanent emergency basis." The major complexes were the X-10 plant operated by Monsanto and then Carbide and Carbide; the K-25 gaseous diffusion separation plant operated by Union Carbide; and the Y-12 separation plant (cyclotron) operated by the Tennessee Eastman Corporation (TEC). Other important areas were the 706-C Laboratory and the carbon burning area. Radioactive elements of concern included P-32, I-131, C-14 and Co-60. It is noted the earliest worker radiation protection level was 0.1 roentgen/day or 100 milliroentgens (mr)/day, although little attention was being paid in the first few years to worker safety. At the level of 100 mr/day there was a definite change in blood chemistry in workers over the short term. This protection level slowly changed, by 1950 was lowered to 15,000 mr/year, and by 1957 was down to 5,000 mr/year or 25 mr/day.[81]

At ORNL, information about production, health, safety and the environment was off limits, and production was of highest priority. Health and environmental information on conditions at ORNL in the 1940s and 1950s was not made available until the 1980s (and a lack of information exists even today). A code of silence prevailed, and security was strictly enforced. Security personnel monitored meetings of more than three persons on the streets, and undercover agents mingled with the public at social occasions. Male workers were recruited under threat of being drafted. Women formed a large part of the workforce especially at Y-12. They were not allowed to ask questions, but to strictly monitor their machines. At Y-12, the cyclotron separation plant, the workers were told to keep the instrument between two prescribed points, and not worry about anything else. At the gaseous diffusion plant K-25, uranium hexafluoride was used, a heavy and corrosive gas in a series of sealed tubes or pipes. One report said that UO_3 was received from Hanford and Savannah River, the UO_3 was converted to UO_2, then UF_4 and UF_6. Much of the work involved maintenance of the massive system of pipes and dealing with leaks of air into the system. Issues were raised by workers as to unsafe conditions, but answers were not given. Workers in production and construction were routinely subject to possible exposure to radiation and chemical substances but never given the opportunity to refuse an assignment or told of the hazards of their work.[81]

Accidents, Human Experiments

In March 1945, an African American cement truck driver- Ebb Cade was involved in a serious traffic accident and placed in the Oak Ridge hospital. While there without his consent, he was injected with 4.7 micrograms of plutonium-239 and monitored for adverse effects. When his teeth and gums became infected, his teeth were extracted and tested for plutonium. He left the hospital on his own power and disappeared. The above case raised critical issues about worker safety of the Manhattan Project during WW2. Before the war, the practice was that those undergoing medical experiments should be given a choice about their participation. The Cade case additionally raised the issue of using humans in experiments, whether race and racism were present, and if some groups of subjects were viewed by the Oak Ridge medical staff as less valuable than others. Some 50 years later in 1995 when the government's Advisory Committee on Human Radiation Experiments issued its report to the public, it was revealed there were many cases like Cade's. Production could not be slowed down or stopped under any circumstances. The later releases of formerly classified documents gave a disturbing picture of human exposure to radiation, chemicals and other hazards.[81]

Serious Worker Radiation Exposure Throughout the Facility in the Early Years

Building 706-C in the X-10 complex aka the Clinton Laboratories was especially hazardous because the lab was experiencing levels of radiation much above its design. The section leader said 706-C was designed for 10 curies of hard gamma and was a one slug U-238 building with a maximum of 10 men. However, it was being used for 400 curies of hard gamma production on a 2300 slug basis and populated by 20 men. Those working with very hot material were receiving radiation overdoses. Furthermore, personal dosimeters and film badges were not being used as prescribed, which underestimated exposure. The Y-12 magnetic separation plant was said to have some of the most severe occupational hazards, especially Departments 185 and 186. Chemical recovery workers faced almost daily hazards from the strong acids used to remove the tiny bits of uranium from various pieces of equipment, and the subsequent processing of the mixture. Radioactive materials, acids and other chemicals were handled in open tanks. These places were also crowded with pipes, hoses and tanks adding to the exposure. Tennessee Eastman Kodak did not do an adequate job of protecting worker safety by admission of its own medical and safety departments. The Chemical Departments, including Departments 185 and 186 with less than 15% of plant workers accounted for 43% of the overall plant accident claims. TEC and ORNL denied a long list of worker injuries in Y-12 and any danger of radiation exposure. TEC showed an extensive history of occupational injury and disability according to documents released many decades later.[81]

An inspection of Department 186 made in February 1945 showed on-going safety problems known by TEC management. Not all employees were wearing safety glasses, fire extinguishers were too heavy to be handled by female workers, respirators were stored in areas containing uranium dust contaminating them before use, lighting and ventilation were unsatisfactory in certain areas- in one case the regular air intake was sealed off because of toxic fumes, ventilation hoods were not being used properly allowing gas to leak into the room, loose mercury was found on the shelves and elsewhere, and one building had not received a copy of the safety rules. In other cases, uranium was escaping into ventilation ducts and onto clothing and the lungs of workers. The carbon burning area of 9201-1 was declared an industrial hygiene hazard requiring immediate attention. The entire facility lacked adequate ventilation. Air filtering equipment had been installed but filters not inserted. In June 1944, TEC realized a critical mass of atomic material was present in Building 9207 of Y-12. TEC wrote to the Army asking for assistance in "stopping reactions resulting from bringing together by accident an unsafe quality of material." Such an accident would have put many lives in immediate harm.[81]

Union Organization, Health and Safety Issues Continue, Army Resistance to Change

Not only did the Manhattan Project Director General Leslie Groves vigorously fight against any limits placed on production but also the establishment of unions at ORNL until around 1946. The Secretary of War was also strongly against unions. But worker dissent and union pressure were growing. At Oak Ridge, worker attempts to influence safety policies were resisted by the Army and government contractors seeking to prove their toughness to AEC to keep their contracts. In 1953, the town council asked AEC to end desegregation in the local schools which the AEC refused to recognize until Brown vs. Board of Education in January 1955. During the Cold War, the AEC continued to avoid responsibility for worker and community health problems as did other corporations dealing with hazardous materials. The Manhattan Project simply "classified" its scientific evidence that radiation and radioactive substances caused cancer. The AEC allowed only limited access to information about radiation dangers and human experimentation. Heavy discussion between non-government physicians and AEC and DOE as regards worker safety and the prevalence of cancer continued well into the 1990s.[81]

Worker Illness Compensation Starting in 2000

In 1999, the Secretary of Energy Bill Richardson admitted that the nuclear weapons industry had harmed its workers through radiation and toxic chemicals, and he said workers would be compensated. Under PL-398, the Energy Employees Occupational Illness Compensation Program Act of 2000, workers were to be compensated $100,000 each for work-related injuries and diseases incurred at atomic facilities including Oak Ridge plus medical care for all work-related health problems. Under PL 107-107, Section 3151, this payment was raised to $150,000 in 2001. A major problem was the program relied upon DOE and contractor records that might no longer exist. Of the 6,900 claims filed for the $150,000 payment in Tennessee up through 2004, 1,609 were approved and 1,444 were denied. This system of compensation was byzantine nationwide, with a very high rejection rate for compensation. Of 17,757 claims made, only 529 complete worker records were located.[81]

Other Research and Development by ONRL

ORNL developed the world's first light water reactor, the ancestor of most modern commercial power stations which was mainly funded by the military for use in nuclear powered submarines and ships of the U.S. Navy. In 1953, portable nuclear reactors were used in remote military bases. ORNL was the only western source of californium-252. In 1964, the Molten Salt Reactor experiment started up and operated until 1969. The High Flux Isotope Reactor was built in 1965 and improved on the work of the X-10 Reactor. Staff in the early 1960s at ORNL was 5,000. In the 1970s, fusion power was created by achieving a plasma temperature of 20 million degrees Kelvin, but these experiments ultimately failed. In 1977, construction began for housing superconducting electromagnets intended to control future fusion reactions. In 1989, the High Flux Reactor was restarted. In 2019, ORNL was producing plutonium for NASA and the space exploration program. Today, ORNL is a very different research center. It has been the site of various supercomputers. The current staff is 5,100.[82]

Environmental Pollution, Start of Waste Cleanup, Extensive Mercury Contamination of Site and Surroundings

In the 1980s, DOE became concerned with pollution surrounding ORNL, and extensive cleanup efforts began. Burial trenches and leaking pipes had caused groundwater contamination, and tanks filled with radioactive waste were sitting still. Costs of remediation at that time were estimated to be in the order of hundreds of millions of dollars. The five oldest reactors on site were deactivated.[82] Massive amounts of waste were said to have been released by the Oak Ridge facility into surrounding air, streams, groundwater and shale formations beneath the plant. Many radioactive elements being released had a half-life of many centuries. The problems of environmental pollution at Oak Ridge continued because of DOE's exemption from federal and state environmental regulations. There were more than 100 different waste discharge pipes from the Y-12 plant alone, and much of the waste from Y-12 was routinely being released untreated in 1983 to the environment after dilution in "treatment ponds." One author said the ORNL was among the most polluted facilities in the DOE system with more than 600 contaminated sites requiring remediation. More than 440,000 cubic meters of low-level waste were disposed of by ORNL through 1991, and 41,000 cubic meters of mixed low-level waste remained on the Reservation in 1991.[81]

Between 1950 and 1963, ORNL's Y-12 plant used 11 million kilograms of mercury for lithium separation. Smaller amounts of mercury were used at other ORNL operations. About 3% of the 11 million kilograms or 738,000 pounds in use was estimated to have been lost to the environment. Some 350,000 pounds of mercury were said to have been released into East Poplar Creek which originated on the Reservation plus releases to other local waters. Although primary mercury discharges from Y-12 essentially stopped in 1963, small amounts of mercury continued to be released from point sources and from contaminated soil and groundwater sources within the Y-12 complex.[77] A 1997 DOE report- "Linking Legacies" admitted mercury was discharged into Oak Ridge waterways since 1950, which remain contaminated. Mercury was released into Oak Ridge soils, and mercury residues remained in pipes and sewers at the facility. This mercury may also still be found at the bottom of the facility's lakes and ponds.[81]

As mercury becomes exposed to streams, soils and organisms, mercury is transformed into the more toxic form of methylmercury which is absorbed by organisms, plant and fish life. Although mercury concentration in East Poplar Creek waters has decreased by 85%, methylmercury levels in waters and fish life have not decreased, and in some cases have increased. Biomagnification has been occurring. Mercury can cause significant neurological harm to animals and humans. Women who eat fish contaminated with mercury have a greatly increased risk of having children with birth defects. ORNL has tried various treatment and recovery methods for reducing mercury in the environment.[83]

9

LOS ALAMOS, NEW MEXICO PLANT

Early History and Background

The Los Alamos National Laboratory (LANL) was established in 1942 for the design of nuclear weapons as part of the Manhattan Project. It comprises 43 square miles and was the assembly site for the world's first atomic bomb. LANL was highly secretive into the late-1990s if not up to the present. It was managed by the University of California. LANL was the heart of the Manhattan Project with billion- dollar budgets and brought together some of the world's most famous scientists, among them numerous Nobel Prize winners. Its first director was J. Robert Oppenheimer. Later because of controversy and changing philosophy, many of the original Los Alamos luminaries chose to leave LANL, and some became outspoken opponents of further development of nuclear weapons. After several reorganizations, LANL is currently managed by Triad National Security.[11,84]

LANL was located on a "secret" mesa 35 miles north of Santa Fe, New Mexico. The LANL facility is shown in Figure 9.1. The mesa was bisected with many canyons in and around the site. The area was protected first by the U.S. Army, later by hired protection services, and always enclosed within the classification system that kept secret a massive amount of information that the Lab scientists had collected over the decades. Around 1998, a 400-strong security force guarded the laboratory together with the community police force. The earliest picture of Los Alamos was that of muddy streets surrounded by barbed wire fences reminding the European refugee scientists who worked there of a concentration camp. The plutonium processing facility-Technical Area TA-55, was possibly one of the most intensely guarded places in the United States. TA-16 was an area where the facility machined and tested explosives. A main objective of the Manhattan Project was the manufacturing and storing of plutonium pits especially at Rocky Flats, Colorado, Los Alamos, Pantex, Texas, and Hanford, Washington. One such plutonium pit is shown in Figure 9.2.[84]

Figure 9.1 The Plutonium Facility at Los Alamos
Source: discover.lanl.gov/publications

Figure 9.2 Plutonium Pit Design from the 1940s, Los Alamos Facility
Credit: Nuclear Watch New Mexico

For Santa Fe citizens located a relatively short distance away, the stolid demeanor of Los Alamos was threatening. Santa Fe and Los Alamos were light-years apart in terms of culture and attitudes, the former with anti-nuclear activists who kept a suspicious watch on their neighbors on the mesa.[11,84]

Plutonium has more explosive power per gram than any other fissionable material. Plutonium-238 is a heat source for sensitive electrical components in satellites, and plutonium-239 is essential in making nuclear weapons. Nearly all the plutonium that exists on the earth today is human made inside nuclear reactors by bombarding a solid uranium mass with neutrons and precipitating out everything but plutonium or element 94. There are various isotopes of plutonium, the prized isotope being plutonium-239. During WW2 there were people at Los Alamos who took risks in recovering plutonium and who received heavy plutonium radiation. The greatest danger is the inhalation of fine particle plutonium which concentrates in the liver and on bone surfaces.

Once it finds its way into the body, it stays there forever, emitting radiation to surrounding tissues. Experts on low-level radiation contend that plutonium spread into the earth's atmosphere by above-ground nuclear tests were responsible for hundreds of thousands of lung cancer deaths as well as genetic abnormalities. The nuclear industry also tells us that everyone born before 1962 has particles of plutonium permanently fixed in their bodies because of fallout from the various atmospheric tests, and several tons of plutonium continue to reside in the biosphere.[11,84]

Sometime after 1990, LANL did extensive on-site assessment of conditions at the facility, and found 770 violations of federal health, safety and environmental standards. A follow-up inspection in 1991 by a DOE team identified 2,300 solid waste disposal sites that needed further study and excavation, with an estimated cleanup cost of $2 billion. The team also found that the Laboratory which was the site of 15 nuclear reactors- one of them among the oldest operating reactors in the country, was not capable of dealing with emergencies. DOE prepared a 1,000-page report describing massive improvements for the Laboratory in order to stay in operation. Deep and disturbing changes were necessary for the thousands of people who worked there. One scientist said the biggest stumbling block was the deeply ingrained culture of the Laboratory. Managers were told to bring in the money and let the scientists play in their sandboxes, which was the attitude of the 1950s and 1960s, extending into the 1970s. Another said... "It's amazing that people find it so hard today to believe that nothing was written down." Progress moving forward was slow and accidents continued to happen.[11]

Despite increasing controversy by a suspicious public, changes in laboratory management, and being forced to deal with large budget shortfalls, by 1997, the Lab would resume underground testing at the Nevada test site conducting sub-critical experiments using small amounts of plutonium that were subject to extreme forces to cause chemical explosions. The purpose was to better understand the effects of aging on weapons in the nuclear stockpile, but activists insisted these tests violated the spirit of the Comprehensive Test Ban. A federal judge refused to halt these tests. One chemist at Los Alamos believed that "weapons-people" practice a kind of compartmentalized thinking that probably protects them from reality. There is always the nagging worry… I hope I am right about this- but if we are wrong, we are responsible for killing millions of persons. Another person said… Weapons designers like other humans are prone to compartmentalized thinking where they assume responsibility for only part of the problem. "At Los Alamos, there was scant guilt over the mission of the Laboratory. If you are against it, you don't work here."[11]

Accidents, Safety and Health, Fires, Criticality, Human Studies Involving Plutonium

Three major nuclear-related accidents were reported to have occurred at LANL. Criticality accidents occurred in August 1945 and May 1946, and a third accident occurred during an annual physical inventory in December 1958. Details were not provided on these three accidents. In November 2008, a drum containing nuclear waste was ruptured according to DOE. A similar accident took place on February 14, 2014, at the Carlsbad Waste Isolation Pilot Plant (WIPP) facility partly caused by LANL mistakes, which resulted in significant disruption and costs across the industry. The Lab was on schedule in 2014 to remove 3,706 cubic meters of hazardous and radioactive wastes stored in above-ground containers with shipment to the WIPP facility near Carlsbad, New Mexico when a lab container ruptured at WIPP, halting operations. LANL was penalized with a $57 million reduction in its 2014 budget for the accident at the WIPP facility. The year 2000 brought danger to the Laboratory in the form of the Cerro Grande Fire which was severe and destroyed several buildings and employees' homes, and forced the Lab to close for two weeks. In August 2011, a near-criticality accident occurred when 8 rods of plutonium were placed too close to each other to take a photo. In the aftermath, 12 of the 14 Lab's safety staff left in anger when their advice was dismissed by management. With lack of safety management, the Plutonium Facility PF-4 was shut down in 2013. In August 2017, the improper storage of plutonium metal could have caused a criticality accident, the incident not being reported by LANL staff.[11,84]

The Manhattan Project was well underway before the effects of plutonium on the human body were known, but scientists of that time were involved or aware of secret work being done and were nervous about it. The first injection of plutonium into humans was carried out at Oak Ridge in April 1945, and tissue samples were examined at LANL. It was reported that some ethical breaches in the conduct of these human radiation experiments would come back and haunt Los Alamos and other AEC/DOE medical facilities. The number of subjects who were given or not given the right of consent and told of the nature of the experiments is highly debatable. A Congressional report much later in 1986 gave details of this research that had been kept secret for years, and certain documents were suppressed to spare the AEC embarrassment and litigation. Total body radiation, radioiodine, tritium and carbon-14 human studies were conducted including those made on very young children and fetuses. Prison inmates in Oregon and Washington were subject to the irradiation of their testicles. Los Alamos was involved in many of these studies. Many volunteers from Los Alamos were direct participants, injected or otherwise.[11]

Unreported Testing, Records Release, Worker Compensation

The above studies and many others were revealed by Secretary of Energy O'Leary at the end of the Cold War. O'Leary said it was time to come clean and disclose as much as possible about the past 50 years of government radiation research including classified information regarding experiments carried out on humans. At the same time, a Congressional investigation revealed that government agencies including LANL had made 13 unannounced releases of radiation into the environment and had not notified the affected populations. A December 15, 1993 GAO report described three tests in 1950 where radioactive lanthanum was released into Bayo Canyon together with small quantities of strontium-90. O'Leary's announcement of DOE's new openness policy was made on December 7, 1993, and caught LANL by surprise causing a crisis at the Lab. Morale in those days was low and questioning the ethics of its research was an added blow. In response, LANL formed a Human Studies Team. Some members asked… how dare they criticize our work? The sheltered community realized it was not getting the respect it once had known. The Team then released to the public more than 1,500 documents amounting to over 21,000 pages of information, much of it as laboratory records and some still in the possession of the researchers themselves. It was discovered certain important items had been thrown away. Such records existed only because individuals had made copies of their work when they retired and stored them at home. LANL claimed they were not involved in the plutonium injection experiments beyond tissue analysis, and in the one case of providing the plutonium solution. The Lab was also confident participants had given their informed consent, although this was not always formally documented.[11]

It was reported many workers in the nuclear field have suffered from radiation. Uranium miners and milling workers have developed lung and kidney cancers among other illnesses. Scientists and other workers exposed to radiation from above-ground testing developed cancers of the lung, thyroid, esophagus, stomach and pancreas, as well as leukemia and other illnesses. Some 107,141 nuclear research workers received greater than $11.6 million in compensation and medical coverage as of July 2015. More than one-quarter of workers filing claims had cancer types recognized by the federal government caused by exposure to radioactive materials. Some 4,900 former Los Alamos workers from WW2 through 2015 have received $555 million for health problems related to their Lab work.[86]

Spies

The most notable spy story was that of Klaus Fuchs. Fuchs was a highly regarded British scientist but also a dedicated communist who worked at Los Alamos early on and a spy for the Soviet Union. It was suggested that Fuchs helped the Soviet Union in reaching their nuclear goal two years earlier than expected. The British government released documents but not until October 1996 concerning Fuchs and Theodore Hall, another agent working for the Russians at Los Alamos. Fuchs was released from prison in 1959 and gave China vital information, but Russia also provided nuclear information to China. Hall was a university professor who retired at Cambridge University and was never arrested. Morris and Lois Cohen were carriers who delivered vital information to the Soviets. Soviet spies had penetrated Los Alamos as early as October 1941. Russia exploded its first atomic bomb in August 1949. The Russian Pavel Sudoplatov report said that other prominent Manhattan project scientists were implicated including J. Robert Oppenheimer, Enrico Fermi, Neils Bohr and Leo Szilard, but the FBI and others found no factual support for the latter accusations.

In later years, David Greenglass- a machinist and George Koval- a health physicist officer were noteworthy spies in the Manhattan Project. Other Los Alamos spies included Harry Gold, Ethel and Julius Greenberg. In 1999, Los Alamos scientist Wen Ho Lee was accused of 59 counts of mishandling classified information by downloading nuclear weapons codes used for computer simulations of nuclear weapons tests onto data tapes, and removing them from the Lab. After 10 months in jail, Lee pleaded guilty to one count and the other 58 were dismissed. Lee had been suspected of having shared U.S. nuclear secrets with China, but investigators could not establish what he did with the downloaded data. Another scientist at Los Alamos, Joan Hilton, moved afterwards to China and said she did not give any secrets to China, but the U.S. believed otherwise. Also, "Persius" at LANL gave the Soviets plans for the hydrogen bomb. General Groves believed the Soviets had obtained very little information, but he was wrong.[6,10,11,84]

Controversy

After WW2, many of the scientists involved in the Manhattan Project went back to their universities. But some were so horrified by what they had created that they devoted the rest of their lives opposing nuclear weapons. Most scientists at Los Alamos continued to believe that having a nuclear stockpile or abundance of bombs was a necessary deterrent to these weapons ever being used. When Secretary of Energy O'Leary ordered DOE to reveal information on plutonium experiments being used on humans- many without their consent, some Los Alamos veterans were disturbed by this. LANL was forced to promote public acceptance and understanding. But they did not excel at this because of the profound secrecy practiced, and LANL lost creditability over this order. One author said… Perhaps these scientists were too arrogant in not telling the public and Congress more about what they were doing with their large and unsupervised annual budgets. Los Alamos researchers complained that neither the press nor the public understood their research, being uneducated in science, and the inability to make emotional judgments about technical issues. What many Los Alamos scientists forgot, for the average person, nuclear physics is very complex and difficult to understand, and provokes fear among the public.[11]

From a peak of 60,000 nuclear weapons, agreements between the U.S. and Russia served to reduce the nuclear arsenal down to 3,500 by the year 2003. Nuclear weapons scientists contend the longer the warheads sit around, the more likely they are likely to undergo deterioration with reduced performance. When on nuclear subs, they move around and also get dropped. It is noted that during the 1990s, scientists at Los Alamos and the Sandia National Laboratory in Albuquerque were working on a mini nuke. Testing is necessary because most of the weapons in the U.S. arsenal have experienced serious reliability problems at one time or another. LANL scientists claim limited underground testing is necessary to determine bomb integrity over time. Senior weapons scientists at LANL do not accept that the nuclear stockpile should be reduced to perhaps 100 or so.[11]

Around August 1995 or slightly before, the Los Alamos community made it clear they would not assume guilt for the Hiroshima and Nagasaki events when a group of Albuquerque schoolchildren approached the Los Alamos County Council for permission to place a Children's Peace Statue in the town. The group had collected more than 40,000 signatures from children across

50 states and 53 nations and raised $20,000 to build the statue. The town, however, believed the statue would be a slap at the Laboratory and attract even more protestors. Parents in town worried about how such a monument would affect their children. The town refused the Children's Peace Statue, which was finally given a home in Albuquerque. However, life-size statues of Oppenheimer and General Groves stand in front of a history wall at the Bradbury Science Museum together with many Cold War weapons artifacts. Peace activists refer to the museum as a public relations tool.[11]

On August 6, 1995- the 50th anniversary of Hiroshima at LANL, a ragged and disheveled protestor had set up a display where he folded sloppy, origami paper cranes and asked anyone who would listen if they knew the story of the thousand cranes. Then a storm came in, and several paper cranes from the display took flight in the wind. The story of a thousand cranes is that of a little girl named Sadako who survived the initial blast of Hiroshima only to succumb 10 years later to leukemia. As she lay sick in the hospital, she dreamed of a Japanese legend that if a sick person folds a thousand cranes, they will get well. She managed to fold over 600 cranes before she died at the age of 12, inspiring her friends to finish the task and collect funds for a statue honoring all the children who suffered from the bomb. The message on the statue reads… "This is our cry, this is our prayer, peace in the world."[11].

Radioactive and Chemical Wastes, Management and Cleanup

At LANL, the laboratory TA-21 which was the ancestor to TA-55, was one of the most heavily contaminated parts of LANL. Tech Area 21 was where early research with plutonium, uranium and other radioactive materials was conducted. Much of the contamination in TA-21 was not cleaned up as of 1998, even though plans had been made in 1996 for removing Buildings 3 and 4 in TA-21 at a cost in the millions of dollars. Building 3 North was a structure that had not been decontaminated since 1945, and water leakage and rodent activity may have been spreading radioactivity. A 1982 LANL study showed pocket gophers burrowing into low-level waste trenches. At Los Alamos, it is said nobody knows where all the stuff is- including underground piping because records were not kept and or were thrown away. LANL scientists said a large chemical plume had formed deep inside the mesa, the source probably being TA-54 where drums of solvents disintegrated and released their contents into soils and rocks. The drums were buried in the 1950s when the practice was to place these loaded drums onto pallets and simply cover them with soil. People in New Mexico have feared plutonium and other elements can be flushed out of the canyons of Pajarito Canyon and find their way to the Rio Grande River into downstream water supplies, but LANL was not convinced. The many canyons running through the Los Alamos nuclear facility are described in Figure 9.3.[11]

Remaining Environmental Radioactivity

"Mixed wastes" represent a substantial part of total waste at nuclear weapons facilities. Exorbitant quantities of these mixed waste must be disposed of because they have been contaminated by contact with radioactive waste including plutonium[11]. As previously cited, around LANL's 43 square mile property, there are more than 2,000 dump sites which have permanently contaminated the environment. LANL has also contributed to thousands of dump sites at 108 other locations in 29 U.S. states. Many receiving streams, whether intermittent or not, in the canyons surrounding or within LANL, have been seriously impacted by LANL's activities over the decades. A small canyon- *Acid Canyon*, below the Los Alamos Nature Center was a dumping ground for the Lab from 1943 to 1964. A chemical waste treatment plant at Los Alamos released more than 30 million gallons of treated and untreated liquid radioactive and chemical wastes laced with tritium, strontium, plutonium and other radioactive materials. Acid Canyon was just one of several canyons used as dumping grounds by the Laboratory.[84,86,87]

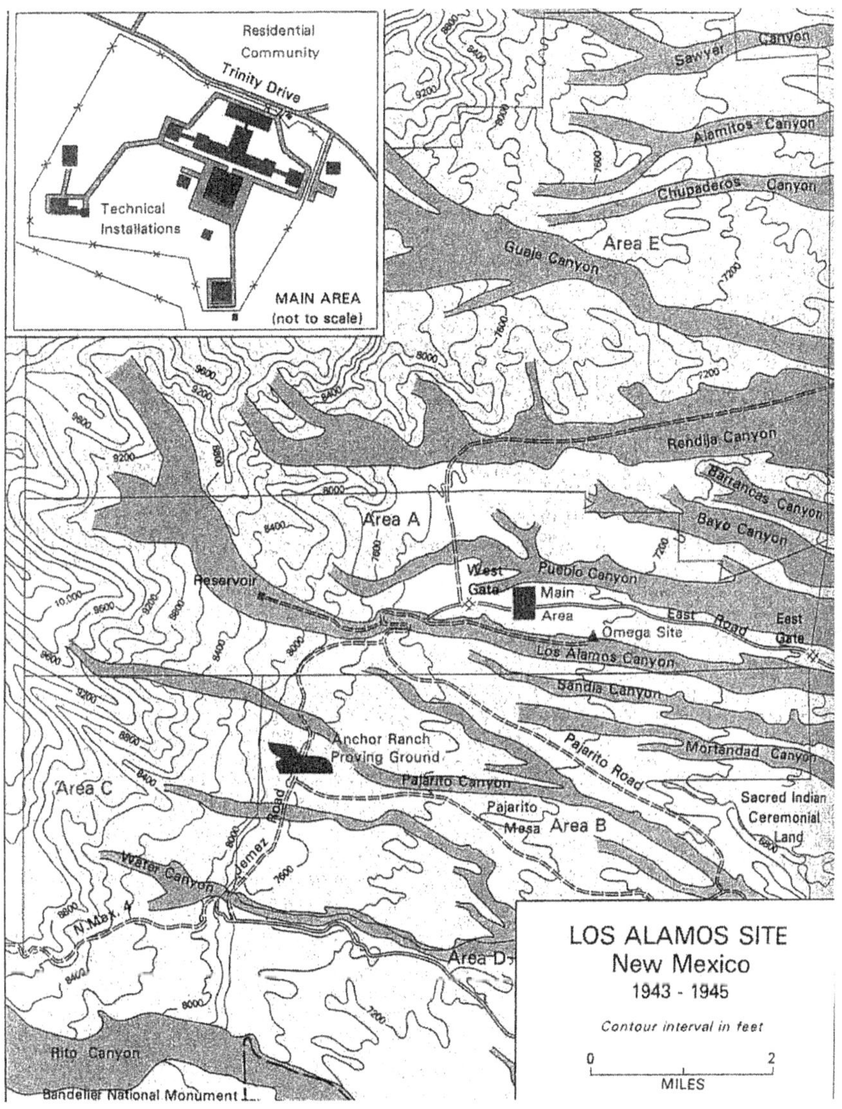

Figure 9.3 Los Alamos Site Map, 1943–1945, Main Area and Canyons

Credit: Wikimedia Commons, Jones-U.S. Army, USDOI

Technical Area-10 and *Bayo Canyon* was the place where radioactive contaminated waste from an "offsite lab" was dumped. This testing area of 350 acres was said to have been cleaned up in the 1960s, leaving an area of 1.5 acres contaminated with strontium-90 some 4 to 8 feet below ground level. The latter area was left intact, but a marker or markers were placed warning of buried radioactive waste, and the area is now used by the public for recreational purposes. Stream gages and electronic monitors sit at the bottom of Acid Canyon to alert the Lab when floodwaters come through that can potentially convey contaminated sediment onward to the Rio Grande River and downstream to a diversion that carries water into Santa Fe's drinking water system.[84,86,87]

At Los Alamos, old 55-gallon drums are still being pulled out of the ground. The wastes dumped in the canyons were buried in unlined trenches and pits and randomly discarded and are now located underneath recreational trails. These sites at Los Alamos included Area G- a 63-acre dumpsite opened in 1957 containing thousands of cubic feet of low-level and mixed transuranic wastes. In 2014, a Lab official said they still had thousands of cubic feet of contaminated waste remaining onsite in 35 pits and 200 shafts at Area G.[84,86,87]

Radioactivity Potential in Stormwater, Large Underground Aquifer, and Surrounding Canyons

A main concern was that Area G is situated directly over the regional groundwater aquifer some 900 to 1,000 feet below the surface. The present status of groundwater quality at LANL was not determined. One official said "Los Alamos will never be clean. It will always have tons of buried waste. Whether the waste is a health hazard is debatable."[84]

The Western Environmental Law Center reported in June 2019 that stormwater from LANL containing PCBs, copper, zinc, nickel and gross alpha radiation in Los Alamos County is threatening public health. Some pollutant levels are said to be more than 10,000 times over public safety limits.

The Law Center on June 28, 2019, said the EPA failed to issue a NPDES permit(s) to control this pollution, and the Law Center notified the EPA of their intent to sue in order to ensure the safety of water supplies in and around LANL. It said the Clean Water Act requires the EPA to regulate stormwater runoff when that runoff is making receiving waters unsafe.[88]

The New Mexico Department (NMED) showed that PCB levels in *Los Alamos Canyon* were more than 11,000 times higher than the New Mexico Human Health water quality criteria and 51 times higher than the New Mexico Wildlife Habitat water quality criteria. PCB levels in *Sandia Canyon* were more than 14,000 times higher than the New Mexico Human Health water quality criteria and 66 times greater than the New Mexico Wildlife Habitat water quality criteria. PCBs in *Pueblo Canyon* were respectively more than 3,500 times higher and 16 times greater than the two water quality criteria.[88]

A State of New Mexico 303d/305b report showed many exceedances of standards for a variety of pollutants at several locations. *Mortandad Canyon* contained high levels of PCBs, mercury, silver, cyanide, copper and gross alpha radiation. *Pajarito Canyon* was impaired for gross alpha, aluminum, PCBs and copper. Urban runoff was said to be the culprit for many of these pollutants. It was cited above that toxic pollution can flow downstream from Los Alamos into the Rio Grande and potentially impact drinking water diversions for both Santa Fe and Albuquerque.[88]

10

PANTEX, TEXAS PLANT

Pantex is the primary U.S. nuclear weapons assembly and disassembly facility for DOE that currently maintains reliability of the U.S. nuclear weapons stockpile. It is located 17 miles northeast of Amarillo, Texas. The plant is managed and operated by Consolidated Nuclear Security (CNS) who also operates the Y-12 plant at Oak Ridge, Tennessee. It is situated on 16,000 acres (25 square miles) of land purchased by the U.S. Army in 1942. As a major national security site, access to the plant and its grounds is strictly controlled and off-limits to all civilians; and the airspace around and above the plant is prohibited to civilian air traffic by the FAA.[89,90]

The plant was authorized in February 1942 as a conventional bomb plant for the U.S. Air Force. It was deactivated when the war ended and remained vacant until 1949 when Texas Tech University purchased the land for $1. In 1951, the AEC exercised a recapture clause and reclaimed the main plant and 10,000 acres of surrounding land for use as a nuclear weapons production facility. The AEC refurbished the plant at a cost of $25 million. In 1989, 6,000 acres of the original site was leased from Texas Tech. In 2010, the plant employed 3,600 persons and had an annual budget of $600 million. In 1994, the plant was listed as a Superfund site with contaminated groundwater being of major concern. Cleanup was reported to be completed in 2010. There have been concerns over the years of increased risk of cancer among Pantex workers.[90]

In 1977, three men were killed in an explosion while machining LX-09, a plastic explosive. In the early 1980s, local Catholic Bishop Leroy Matthiessen tried to persuade Catholic workers at the plant to leave their jobs, offering financial support to those who would do so. In 2005, the Project on Government Oversight claimed that Pantex workers could have been responsible for a nuclear explosion by applying too much pressure on an obsolete W56 warhead in the process of dismantling it. A catastrophe was narrowly avoided.[90]

After the Cold War, there was a great surplus of plutonium, with the world's stockpile being more than 1,200 tons, and an additional 75 tons per year being added annually from commercial reactors. This commercial output was expected soon to be greater than that coming from defense installations. The Los Alamos National Lab was involved in the question of available storage. Plutonium pits, the major component of nuclear weapons, were of major concern. Thousands of these pits had been produced at the Rocky Flats facility near Denver from 1953 to 1989. When Rocky Flats closed, after 1989, reserve and surplus pits along with pits recovered from disassembled nuclear weapons totaling over 12,000 pieces were stored at Pantex. Plutonium pit production was relocated to the Los Alamos National Laboratory in New Mexico in 1996.[11,74]

Besides the storage of other radioactive materials, the main question at Pantex was how to safely store the plutonium pits without achieving critical mass. Scientists knew when plutonium pits aged, they became increasingly toxic to the workers who had to remove them from decommissioned nuclear weapons. Metallic plutonium, notably in the form of the plutonium-gallium alloy, degrades by corrosion and self-irradiation. The plutonium changes to americium, a strong beta emitter that causes a 25-year-old pit to be five times more radioactive than the original pit. The dismantling of a nuclear weapon at Pantex involves removing all the sub-systems of the warhead: the high explosives in the primary stage, the fission device that triggers the explosion of the secondary stage or fusion component of the weapon, and the chemical explosives that cradle the secondary.

The "interim procedure" at Pantex is to store the pits in large, shielded bunkers until a better solution can be found. Figure 10.1 shows some of the storage bunkers. Ironically, Pantex had been the site where most of the 50,000 U.S. nuclear weapons had been originally assembled. In the final dismantling steps, components are crushed to disguise their precise purpose, and useful materials extracted and recycled. Over 800 nuclear weapons were dismantled in FY 1995 and over 1,000 were scheduled for dismantling in FY 1996. Some of the weapons being handled are shown in Figures 10.2 and 10.3.[11,74]

Figure 10.1 Pantex Facility, Radioactive Materials in Shielded Bunkers, November 1957
Credit: Wikimedia Commons and USDOE

Figure 10.2 Pantex Facility, B-61 Gravity Bomb being Disassembled.

Credit: Wikimedia Commons and U.S. Navy

Figure 10.3 Pantex Facility, B53 Nuclear Bomb being Handled, February 2011

Credit: Wikimedia Commons and NASA

Residents of Amarillo, Texas have opposed long-term storage of plutonium triggers for fear of an accident and contamination of the underground Ogalala Reservoir. As of 1998 or before, there were over 7,000 pits and a large number of complete nuclear weapons stored in "igloos" at Pantex. For the task of more safely removing and storing aging pits, a robotic shuttle called ARIES, moves the plutonium pit into a glove box, slices it open, and reduces it to powder by exposing it to nitrogen gas. The americium is stripped away, and plutonium melted into an ingot, then sealed in a can. The ARIES system can be moved from site to site and has even been loaned to the former Soviet Union. An accident at LANL in November 1993 showed the weakness of short-term storage. A can of plutonium inside a concrete bunker at LANL swelled and burst open. Six such events were reported in the ten previous years.[11,71]

A recent report of November 2022 said that Pantex had at least 20,000 plutonium pits, each roughly the size of a grapefruit in storage. Previously this was called "interim storage" with the pits stored inside shielded bunkers. Since plutonium is known to be deadly to humans in very tiny amounts, this complicates their safekeeping. Pits recycled from B53 bombs enter the bunkers which are already overcrowded and overtaxed. Torrential rains in 2010 and 2017 flooded a major plutonium storage area at the Pantex site, and the repairs cost hundreds of millions of dollars. Previous U.S. Presidents made plans to get rid of excessive plutonium stocks, but no strategy so far has succeeded.[91]

11

SAVANNAH RIVER, SOUTH CAROLINA PLANT

The Savannah River nuclear facility is located about 25 miles southeast of Augusta, Georgia. Savannah River is an operating non-commercial nuclear site previously under the Manhattan Project, in contrast to all others which are closed and in partial or total cleanup. It is known there were numerous transfers over the years of radioactive material and waste between this site and major facilities in the Manhattan project network including Los Alamos, Rocky Flats and others. Savannah River was built in the early 1950s to refine nuclear materials for reuse in nuclear weapons. The site covers 310 square miles and employs more than 10,000 people. In the 1990s, more than 20,000 people were employed at the plant. Savannah River had five major reactors named the R, P, K, L and C reactors, started up between December 1953 and March 1955. None of these reactors are currently operational, although two of the reactor buildings are used to store nuclear materials. Figure 11.1 shows P and R reactors under construction at Savannah River. This facility, which is also close to Aiken, South Carolina was said to be one of the most contaminated nuclear sites on earth.[12,13,92]

Figure 11.1 Savannah River Facility, Construction of P & R Reactors
Credit: Wikimedia Commons and energy.gov

The Savannah River facility is the only currently operating radiochemical separation facility and tritium facility in the United States. The plant mainly produces tritium and plutonium. Tritium is an essential component in nuclear weapons. Construction of mixed oxide fuel manufacturing was underway but halted in 2019 before its completion. The latter operation was intended to convert legacy weapons-grade plutonium into fuel suitable for commercial nuclear reactors.[12,13,92]

F Canyon- The world's first operational PUREX separation plant began operating in November 1954. PUREX extracted plutonium and uranium from materials irradiated in the reactors. **H Canyon**, another chemical separation facility came online later, followed by the tritium facility. In 1962, the Heavy Water Components Test Reactor (HWCTR) went into operation, advancing the heavy water system for commercial use. However, this system was shut down in 1964. In 1963, the first shipment of offsite spent reactor fuels was received at Savannah River. Californium-252 and cirium-244 used as a heat source in space exploration were produced. In 1977, the Plutonium Fuel Fabrication Facility (PUFF) was started up. In 1996, F Canyon separation was restarted.[92]

The HB-line was started up in 1985 producing plutonium-238 for NASA's deep-space program. Construction of the Saltstone and the Replacement Tritium facility started in 1986. In 1991, the M-Area receiving basin for spent fuels was shut down. The K Reactor was shut down in 1992, placed in cold storage, and shut down once again in 1996. In 1993-1994, construction began on the Consolidation Incineration facility, tritium was introduced into the replacement Tritium facility, and radioactive operations were resumed.[92]

In 2000, Savannah River was selected as the location for three new plutonium activities: MOX Fuel Fabrication; plutonium pit disassembly and conversion (likely replacing previous operations at Rocky Flats); and plutonium immobilization. The MOX Fuel Fabrication facility was created to satisfy the nuclear non-proliferation agreement between Russia and the U.S. by processing plutonium into MOX fuel. Russia completed its assignment in 2014-2015 but in the U.S., this task will take decades, and costs are reported to be excessive.

In 2001, some 10,000 drums of transuranic waste were shipped from Savannah River to the Waste Isolation Pilot Plant (WIPP) in New Mexico. In 2005, Savannah River's first shipment of neptunium oxide arrived at the Argonne West Laboratory in Idaho. In 2013, the remainder of depleted uranium metal was shipped from M Area to Envirocare of Utah.[12,92]

In the cleanup of radiological and chemical wastes at the Savannah River facility, the groundwater remediation and vitrification programs stand out, but information in this area is lacking. In 1981, the environmental cleanup program began with settling basin cleanup in the M Area. A full-scale ground-water remediation system was constructed in the M Area, followed by construction of the "Effluent Treatment Project" in 1986 for treating low-level radioactive wastewater from the F Area and H Area separation facilities. In the 1990s, a tank farm was present consisting of 50 underground tanks each as big as the Capitol Dome containing high-level plutonium wastes. In 1996, radioactive materials were subject to the vitrification process, and in May 2019, 35 million gallons of high-level radioactive waste contained in 43 tanks were reported "vitrified." The product was intended for final disposal at Yucca Mountain in Nevada, but it still remains in South Carolina. The Savannah River facility is "vulnerable" if cleanup does not proceed as planned. It is noted the vitrification process was delayed at the Hanford, Washington nuclear facility because of certain operational problems, and did not succeed at the Rocky Flats facility.[12,13,92]

The Aiken County Chamber of Commerce of South Carolina filed a lawsuit against the federal government claiming they have become a dumping ground for unprocessed weapons grade plutonium for the indefinite future. The federal government filed for dismissal, which was granted in February 2017. The state of South Carolina similarly sued the federal government claiming DOE had not prepared an environmental impact statement concerning the long-term storage of plutonium in the state, and additionally that the government had failed to follow statutory provisions in obtaining a waiver to cease construction at the facility. In January 2019, the Circuit Court of Appeals rejected the state's suit. In October 2019, the U.S. Supreme Court rejected the state's petition, allowing the lower court's ruling to stand.[92] Information on past and ongoing operations and cleanup programs at the Savannah River nuclear plant is considerably incomplete.

12

HANFORD, WASHINGTON PLANT

Early History of Hanford

The Corps of Engineers under the Manhattan Project made its first reconnaissance for selecting a plutonium processing plant in December 1942. The first plutonium nuclear reactor was built at Hanford in the south-central part of the state of Washington, and it produced the material for the first atomic bomb explosion near Alamogordo, New Mexico on July 16, 1945, aka the Trinity test atomic bomb. In the Hanford region, some 1,500 persons had to leave their family lands and homes after being evicted by the federal government, and for this they were offered bottom dollar. The Nez Perce Native American tribe who was badly defeated by the U.S. Army in 1877, were the first to be displaced. General Leslie Groves arranged with the Secretary of War Harry Simpson to seize their lands under the War Powers Act. The federal government eventually acquired more than 6,000 square miles of property. A regional location map of the Hanford Reservation in southeastern

Washington state is given in Figure 12.1, together with site information. Individual areas within the Reservation are shown in Figure 12.2[1,6,93]

The Hanford Site

- Originally 640 square miles – now 586 miles

- Established as part of the Manhattan Engineering District – defense mission

- 1943 – 1,500 people in residence 1944 – 51,000 people in residence

- Currently 8,700 DOE, prime and pre-selected subcontractor employees (excludes many)

  - Hanford - $2.03 B/annual budget

- Currently 4,500 employees at PNNL

  - $1B/annual budget

- From defense mission to environmental in late 1980's – now largest environmental cleanup in the world

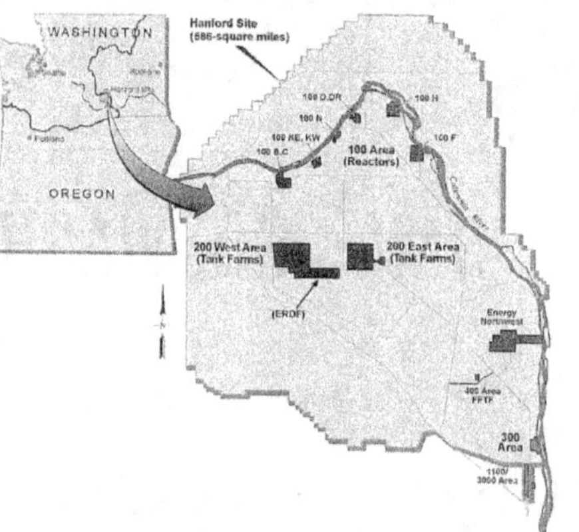

Figure 12.1 Hanford Facility, Location Map and Plant Facts, May 2013
Credit: USDOE, Hanford Site Overview

The reactors were built in a 100-acre area at the far northern end of the site known as the 100 Area. South of the reactors were located the plants for separating plutonium from the irradiated uranium, known as the 200 Area. The safer facilities for producing uranium fuel and conducting tests were located north of the employees' homes and known as the 300 Area at the SE corner of the reservation. The total Reservation was bordered on its north and east sides by the Columbia River. The Corps requested that all regional newspapers keep their activities secret, but the story leaked out in April 1943 because hundreds of displaced people reported they were receiving worse treatment than the "Japs". For some three and one-half years, the word "Hanford" virtually disappeared from the newspapers of the Pacific Northwest.[1,6,93]

Hanford/PNNL Sites

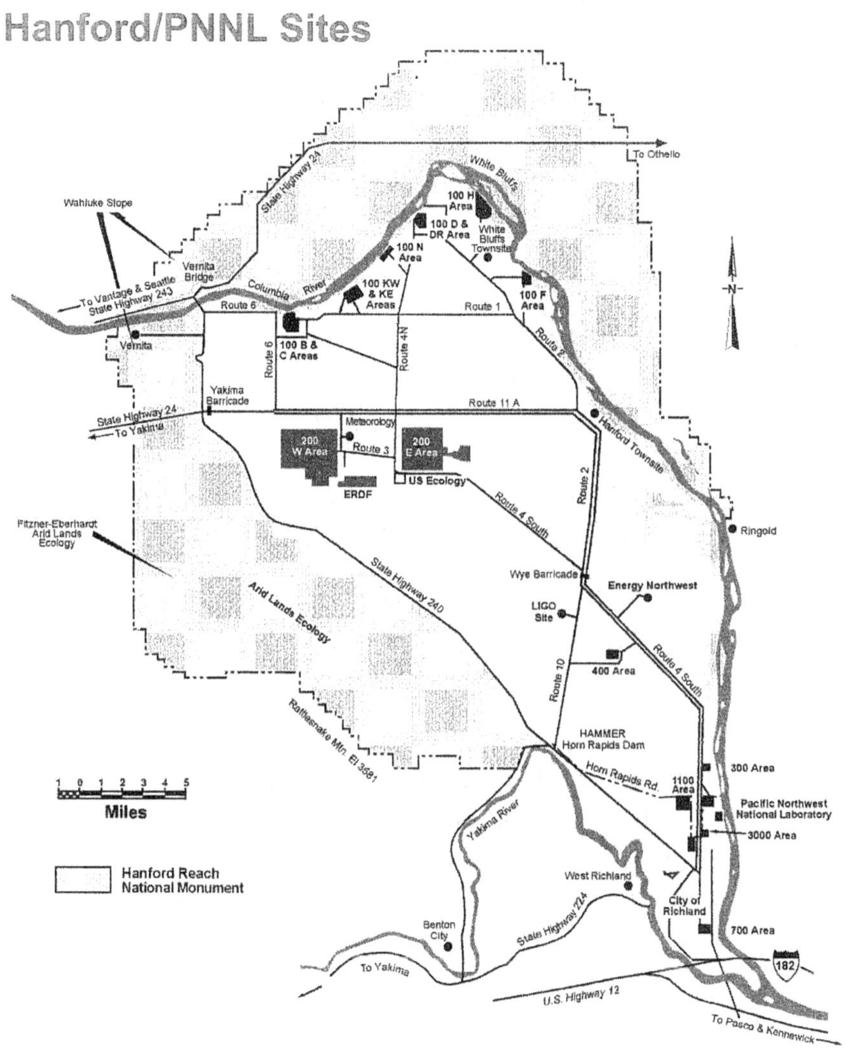

Figure 12.2 Hanford Facility, Immediate Site Map, May 2013

Credit: USDOE, Hanford Site Overview

B Reactor Followed by Reactors A, C Through F

Construction at Hanford was well underway in 1943-1944, and the peak workforce reached 45,000 in May 1944. Huge barracks were erected together with hundreds of hutments and trailer camps. Despite manpower shortages, the Corps and DuPont initially resisted hiring African Americans, but eventually they were accepted and put into segregated housing. Living conditions were primitive, and intoxication and prostitution were common. In late-1943, Hanford's first reactor was under construction named the B-Reactor. It was the first large-scale plutonium production reactor in the world. Cooling water was pumped through aluminum tubes at the rate of 30,000 gallons per minute (gpm). Six reactor sites were laid out labeled A through F, strung out along the banks of the Columbia River, although only three were needed to produce plutonium for the Manhattan Project. The T, B and U separation plants were located about 5 miles south of the reactors.[1,93]

The B Reactor had aluminum tubes inserted into its graphite blocks, the tubes serving as fuel elements containing uranium called "slugs". The U-235 atoms in the slugs would start fissioning inside the reactor, and the neutrons from the fissions kept the chain reaction going for conversion of U-238 into U-239. Fresh uranium would be inserted into the tubes, causing the cooked fuel to leave the back of the reactor and enter a pool of water. After about two weeks to allow the U-239 to decay into neptunium-239 and thereafter plutonium-239, the fuel elements were sent to the separation plants to have plutonium-239 separated from the uranium. Immense amounts of heat were generated inside the reactor. Cooling water which was Columbia River water entered the aluminum tubes to keep the fuel from melting. Some 2,000 tubes were used in the reactor. Water cooled rods also served to keep the chain reaction under control.[1]

It was quickly found the uranium slugs could not be put into the reactor as bare metal but needed to be coated. Otherwise, the uranium eroded, and the spent cooling water carried highly radioactive fission products back to the Columbia River. Experiments indicated the slugs could be immersed in aluminum baths operating at unexpectedly low temperatures to enable coating of the slugs. Also, the reactor was being poisoned by xenon-135 which was absorbing needed neutrons. Xenon-135 was just one of the 30 radioactive isotopes of xenon. This problem was overcome, and by the end of 1945, all three Hanford reactors were making plutonium. After spending several weeks being bombarded by neutrons, the fuel elements deposited in the pool behind Reactor B contained a wealth of radioactive fission products. Some of these decayed rapidly, but the longer-lived ones like strontium-90 and cesium-137 remained dangerous for a very long time. Operators used long tongs to reach to the bottom of the pool and then placed these materials into buckets. They maneuvered the buckets into lead-lined casks, raised the casks up, and placed them on flatbed railroad cars which traveled 5 miles south to immense concrete monoliths which were the three separation plants.[1]

Plutonium Separation Plants

The separation plants were gray cuboid structures nearly 900 feet long and 65 feet wide with no windows and few doors, aka concrete canyons. Down the long side of each building ran about 40 cells each with thick concrete walls, topped by a 35-ton cover, and containing different equipment from the other cells. In December 1944, the first irradiated fuel elements arrived at the T separations plant. The operator of an overhead crane unloaded the slugs which the workers called "lags" after they were irradiated, and placed them into a very large steel tank. The tank was then filled with sodium hydroxide which dissolved the cladding around the uranium metal. This resulted in about 3,000 gallons of liquid feed and dissolved in this feed was less than a tablespoon of plutonium, which had to separated out from everything else. The materials proceeded down the building through the various cells which had chemicals added to them causing the uranium and fission products to become solids whereas the plutonium remained a liquid. Large centrifuges spinning at 15 to 30 rpm moved solids to the outside of the bowl while liquids stayed in the center. Operators at a certain point added chemicals which turned the plutonium into a solid, which was separated out in another centrifuge. Moving down the line, the plutonium became more concentrated.

All of this was done by remote control. Crane operators became experts at lifting the 5-foot-thick concrete lids off the cells and remotely changing pipe fittings. No pumps were used to move materials from one cell to the next; instead steam ejectors were used. Eventually there were five separation plants. The ultrapure plutonium went to an adjoining finishing shop which prepared this material to be shipped offsite.[1]

The making of plutonium at Hanford generated tremendously large amounts of gaseous, liquid and solid waste. It was reported after a new batch of fuel elements arrived at the separation plants that "great plumes of brown fumes blossomed above the concrete canyons, climbing thousands of feet into the air, which drifted sideways as they cooled, blown by winds aloft." These plumes consisted partly of mist coming from the nitric acid used to dissolve the uranium, but other more ominous materials went up the stacks. Some of the fission products were in the form of gases, like xenon. These elements got swept up in the exhaust air coming out of the canyon buildings, and included radioactive isotopes of iodine, strontium, ruthenium, and cerium. Hanford tried to release radioactive gases only when there was a strong wind for widespread dispersion. However, Hanford sat in a low basin, and when winds were calm, the contaminated air remained low to the ground. When the wind was blowing, plumes of radioactive gases crested the surrounding hills and headed to downwind towns.[1]

Spent Cooling Waters, Releases to Columbia River and Groundwater

Another waste was the spent cooling waters from the reactors returning back to the Columbia River. In the two seconds it took for Columbia River water to flow through the reactor's process tubes, the neutron flux intensity could make almost any natural element radioactive including calcium, chromium, zinc, etc. If there were leaks in the process tubes, uranium and its fission products entered the water. The spent water was held in holding basins for a matter of hours, allowing short-lived elements to decay, but longer-lived radioisotopes were sent to the Columbia River. Up to 75,000 gpm were diverted from the Columbia River to cool the reactors. Fifty miles of Columbia River flowed adjacent to the Hanford reservation. With the early plant, the reactors were being cleaned frequently, and the chemical burden on the Columbia River was said to be greater than the radiological burden. The heat burden on the Columbia River was so severe, that in 1962, an agreement was made between DOE and the Department of Interior whereby additional cold water was taken from Lake Roosevelt behind the Grand Coulee Dam, so that the reactors could continue functioning with cool water. It was reported this was a possible reason for a later shift in plutonium manufacturing to the Savannah, South Carolina nuclear plant.[1,93,94]

A U.S. government report released in 1992 estimated that 685,000 curies of radioactive iodine had been released to the Columbia River between 1944 and 1947. Another report cited 8,000 curies going to the Columbia River daily. Hanford knew as early as 1946 that radioactivity was going up the ecological food chain. Columbia River fish were contaminated, an impact felt disproportionately by Native Americans who depended upon the river for their usual diets. Radiation was later measured in the Columbia River 200 miles downstream as far west as the Washington and Oregon Pacific coasts. Radioactive wastewater also flowed from the fuel fabrication works and the separation plants. Originally, these wastes were released onto the plant grounds which turned into muddy swamps which then became increasingly radioactive. When these swamps dried in warmer weather, radioactive dust was spread by the winds across the landscape. Measures were taken to force this water underground, which led to other adverse consequences. Groundwater mounding was seen below certain buildings and other areas making the tracing and movement of groundwater more difficult.[1,17,93,94]

Offsite Radiation, Underground Waste Storage, Worker Safety

Livestock, principally milk-giving cows, around and downwind of Hanford were said to be eating irradiated crops and the people were drinking the milk and eating the grains, fruits and vegetables that were similarly impacted.

For a while, the population was told to get their milk from sources west of Hanford. Certain species of grass were absorbing more radiation, and the government told the population to cut down these grasses. There was a debate whether certain fishing areas needed to be closed, and surveys were conducted of the public as to how many fish they consumed weekly. Concerns over increased cancer was spreading, especially with I-131 in the thyroid and GI tract cancer. But this was before widespread environmental studies were conducted.[1,93,94]

Huge amounts of solid waste were also generated: boots, gloves, worn-out reactor parts, tools, cardboard, soil, glassware, wire, plastics, radioactive dead animals- enough waste to fill thousands of railroad boxcars, and the boxcars themselves became radioactive solid waste. This waste went into hundreds of landfills, trenches, tunnels and dumps across the site. However, when rain fell, it percolated through the soil covering the waste and going through the waste itself, carrying radioactivity deeper into the soils and groundwater. There was another waste which was worse. The separation plants used very large quantities of chemicals. Once they were used, they were too radioactive for reuse. These radioactive chemicals were carried through pipes to gigantic underground tanks near the separation plants. 177 tanks were built at Hanford to hold high-level radioactive wastes, each tank as large as an auditorium. Workers wore radiation monitors, but they were not told they were working around radiation. Safety inspectors and health personnel used special codes for describing objects and actions dealing with radioactivity rather than citing the actual word itself. Plutonium inside the human body was very dangerous, but production had priority over safety.[1,93]

Plutonium Transport to Los Alamos to be Made into an Atomic Bomb

On February 3, 1945, the first shipment of plutonium from Hanford arrived at Los Alamos, the center for designing nuclear weapons. Because of criticality issues, the Los Alamos scientists had to design an implosion situation, i.e., an inward explosion, to get plutonium to work inside the bomb, which was a difficult assignment at best. When detonators on the outside of the sphere simultaneously ignited their explosives, a shock wave traveling inward was designed to compress a small sphere of plutonium at the bomb's core. If implosion could be perfected, it would be possible for the U.S. to have an almost unlimited supply of atomic bombs and allow for their use after the war. Regardless of whether the Germans had the bomb or not, the message was out if the U.S. and its allies had such a weapon, they were going to use it. In March 1945, Groves was urging Hanford to produce more plutonium and more bombs and taking various shortcuts to do so. However, this speed-up had a severe environmental toll. Chemical waste increased in quantity. Less days for fissionable material in the ponds gave less time for radioactive materials to decay, and release of radioactive elements into the water and air was substantially increased. Transport of plutonium to Los Alamos was made in truck convoys and then flown to Santa Fe in C-47 cargo planes, the public and transport workers largely unprotected.[1]

By the end of 1945, Groves kept the three Hanford reactors going, and expected to have 20 Nagasaki-plutonium implosion-type bombs in the U.S. nuclear arsenal. Groves continued to maintain utter secrecy at Hanford and the other nuclear facilities. On August 1, 1946, Truman signed the Atomic Energy Act which took control of what had been the Manhattan Project, which called for a massive expansion of America's nuclear capabilities, and guaranteed the military would have a powerful influence on nuclear affairs. The AEC was created shortly afterwards. Groves in a confidential letter, "argued if there are to be atomic bombs in the world, we must have the best, the biggest and the most." But many U.S. scientists and others were calling for international control of this weapon of mass destruction.[1]

Building The Nuclear Arsenal, Expansion of Hanford Operations, K-West, K-East, and the N Reactors

The AEC knew that military needs superseded civilian needs, they wanted what Hanford could produce, and the race was on once again. The mood in the U.S. was somber with the onset of the Cold War. It was then decided that the original three reactors (B, D and F) were not enough. The H Reactor located between the D and F Reactors became operational in October 1949; the C Reactor was built next to the B Reactor; and the DR Reactor was built next to the D Reactor which was having problems. A view of the Hanford D and DR Reactors as of 1955 is shown by Figure 12.3.[1,93]

Figure 12.3 Hanford D and DR Reactors in 1955
Credit: AEC and Hanford Nuclear Book/Revelator

Compared to the original three reactors, the newer reactors operated at higher power levels which meant that more cooling water had to pass through the process tubes resulting in more radioactivity in the spent water flow back to the Columbia River. To keep up, a fourth gigantic canyon building using a highly flammable solvent to extract plutonium and uranium out of the irradiated fuel elements was built, aka the REDOX Plant. This plant was more efficient than the previous three separation plants but generated even more toxic and radioactive wastes. General Electric (GE) dumped these wastes into the soil alongside the canyon buildings. For the most toxic and radioactive chemical wastes, GE built 30 very large single-shell tanks and started to fill them up with waste.[1]

By the 1950s, the REDOX Plant was not sufficient to meet the plutonium demands and a fifth canyon building was erected aka the PUREX Plant which went online in January 1956. More single-shell tanks were built to hold the wastes until one could determine what to do with them. In 1952, the AEC announced that GE would build two new reactors known as the K-West and K-East Reactors, which were to be much larger than the first 6 reactors, and operating at energy levels almost 8 times the level of Reactor B. In 1953, U.S. President Eisenhower and Soviet leader Nikita Khrushchev, were overseeing enormous expansions of their nuclear arsenals. Both Hanford and the Mayak plant in the Soviet Union maintained frenetic plutonium production schedules. Security at Hanford was incredibly tight. Levels of plutonium production were top secret which meant that the amounts of radioactivity being released into the air and water also had to remain secret, and production had priority over everything else.[1,94]

In 1966, the N Reactor- the ninth and final reactor at Hanford- was brought online and was the largest power reactor in the world. The fears of many atomic scientists were realized when the U.S. had more than 30,000 nuclear warheads that it could use against the Soviet Union and its allies. The Soviets had far fewer- about 6,000, but it was increasing its numbers rapidly. The total number of warheads in the world soon exceeded 65,000 constituting one million times the destructive power of the bomb dropped on Nagasaki. These warheads are stacked today in the depots of military bases around the world, ready to be dropped by various long-range bombers. They are found as intercontinental missiles in underground silos and in submarines cruising the world's oceans. In 1965 the situation was described as MAD-mutually assured destruction.[1]

The Beginning of the End for Hanford, The White Train

By 1968, the reactors at Hanford and Savannah River had produced 90 metric tons of weapon-grade plutonium- equivalent to more than 14,500 Nagasaki-type bombs. But this was the time of the Comprehensive Test Ban. At Hanford, the first reactor to close was the replacement for the D Reactor in 1964. The H and F Reactors closed in 1965 followed by the D, B and C reactors in the few years thereafter. The K-West and K-East Reactors closed in 1971, and the lone operating reactor was the N Reactor.

Although not producing plutonium since 1966, the N Reactor was upgraded in 1982, and started producing plutonium once again. In 1983, the PUREX separation plant, after a 11-year shutdown, was back in business. Also, the experimental reactor- the Fast Flux facility, began operating in 1980. But in 1982, the nuclear weapons freeze campaign hit the nation. The World Citizens for Peace became prominent, and one of its main targets was the White Train which carried nuclear weapons from the Pantex assembly plant in Texas to military bases around the country. One destination was the Bangor submarine base 20 miles west of Seattle, perhaps the largest concentration of nuclear weapons anywhere in the U.S. On March 21, 1983, World Citizens for Peace held a midnight vigil and protest at the Pasco depot close to Hanford, as a train thought to be carrying 100 nuclear weapons passed by. After that, the federal government used unmarked tractor-trailers to move the warheads. Later, World Citizens for Peace held a demonstration at Hanford's main gate to protest the startup of the PUREX Plant. This followed the Three Mile nuclear accident in Pennsylvania in March 1979.[1]

By 1979, the environmental movement had turned against nuclear power. In 1981, the Washington Public Power Supply System halted construction on 2 of the 5 reactors it was building- one of which was on the Hanford reservation. In 1983, it mothballed 2 half-finished reactors. And later that year, it defaulted on billions of dollars in bonds previously sold for the promise of future electricity. Only one of its 5 new reactors was finished. In 1986, there was the Chernobyl nuclear power plant disaster in northern

Ukraine. Chernobyl was designed to produce both electricity and plutonium for weapons. The N Reactor at Hanford had a reactor core design like that of Chernobyl. A few months later, the N Reactor was shut down mainly because it lacked a containment vessel, and it never operated again. The problems posed by Hanford's wastes were becoming impossible to ignore.[1]

Waste Inventory and Cleanup, Problems with Large Underground Storage Tanks

By 1983, Hanford had built 177 huge underground tanks near the separation plants originally with only a single liner, and later with two layers of steel linings. These tanks ranging in size from 55,400 gallons to 1 million gallons in capacity each, contained a total of 53 million gallons of chemical and radiological high-level wastes from forty years of plutonium production. If held at arms-length, this waste would deliver a fatal human dose in just a few minutes. These tank farms for Hanford are shown by Figure 12.4 adjoining the waste treatment plants, which in turn are shown in additional detail in Figure 12.5. The initial storage tanks at Hanford had an estimated life of 20 years, but 75 years later, the wastes continued to sit in these tanks. DOE seemed to lack information as to the extent that 27 of the double-shell tanks might be subject to corrosion. Also, some of the "empty" single-lined tanks continued to leak. They were buried under 6 to 11 feet of dirt, and the tanks were always meant to be an interim solution for liquid waste from the separation plants, but nobody had come up with a permanent solution. The early single-shell tanks began to leak not long after they were built, and one million gallons of waste from up to 69 tanks had seeped downward. In 2011, 149 tanks were "interim stabilized" by pumping nearly all of the liquid waste into 28 newer double-shell tanks. In the process, solids and debris remaining in the tanks were left in place. Later, DOE discovered at least 14 single-shell tanks were intaking water and one was still leaking waste into groundwater. Further, there was leakage from one double-lined tank. Public interest groups believed contamination had escaped into the groundwater that fed the Columbia River, and a plume of cesium also formed under the tank farm.[1,11,93,97,98]

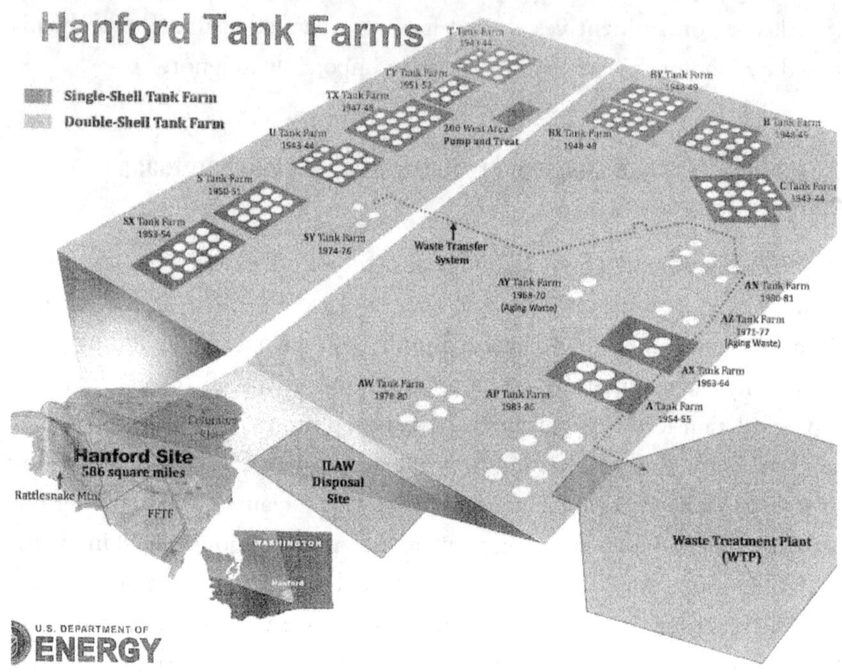

Figure 12.4 Hanford Single and Double Shell Tank Farms, May 2013
Credit: USDOE, Hanford Site Overview

Groundwater Contamination

In 2014, DOE announced further delays in construction of the wastewater treatment plant and immobilization plant intended to receive the tank farm wastes. The treatment plant was located directly east of the tank farm area. In addition to high-level wastes, more than 400 billion gallons of spent water/wastewater from Hanford's reactors, canyon buildings and other facilities had been released directly into the ground, more than one-thousand times the volume in the huge high-level waste tanks. Eventually some of this contamination entered the groundwater table and was moving toward the Columbia River. Hanford continued to be the most contaminated nuclear site in the U.S. Also, environmentalists were saying that radiation released from nuclear facilities was making people sick. But the local people rallied in support of the Hanford facility, and the link between Hanford and health effects was never clear-cut.[1,93,95]

Waste Treatment Plant

Figure 12.5 Hanford Waste Treatment Plants, May 2013
Credit: USDOE, Hanford Site Overview

Release of Documents, The "Green Run."

In 1986, the controversy over possible health effects ignited anew. In February 1986, DOE released 19,000 pages of documents describing the history of Hanford's operations. The papers were reviewed to ensure they would not reveal any secrets or cast Hanford in a bad light. Nevertheless, the documents showed Hanford had released far more radioactivity into the air, water and soil than previously made public. The early Hanford plant used only filters to screen out gaseous emissions, but they did not work well. One major example of information withheld was the "Green Run". In December 1949, operators at the T separation plant deliberately dissolved a ton of irradiated fuel elements only 14 days after leaving the reactors. Radioactive iodine and xenon from the fuel elements spread into the atmosphere together with planes and ground-based tracking of the radioactive cloud. The release was intentional-to determine how much plutonium the Soviet Union was processing assuming the Soviet equivalent of Hanford was dissolving "green" fuel to produce a bomb as soon as possible. But on the day of the Green Run, it rained and the winds shifted erratically making the experiment dubious, with radioactive iodine falling on plants, livestock, and persons in eastern Washington state. The Green Run was estimated to have released 8,000 curies of I-131 over a period of many days and represented one of the largest radiological releases in U.S. history.[1,16,93,94]

Lawsuits, More Groundwater Contamination

In 1987, the DOE released a second batch of documents, and the media was full of stories about Hanford's radioactive releases. Some years later, thousands of lawsuits were filed by persons seeking damages said to be caused by Hanford's activities. Damages were in the millions of dollars. On August 17 and 21, 2017, hundreds if not thousands of gallons of rainwater that had come into contact with the contents of contaminated boxes were dumped directly onto the ground without a permit and state notification. This was related verbally and in documents given to local media.[1,93,94,99]

Collapse of Tunnel Holding Rail Cars with Radioactive Waste

On May 9, 2017, a portion of an underground tunnel containing rail cars filled with radioactive waste collapsed next to the PUREX Plant in the 200 East Area forcing evacuation of persons near the site, and hundreds of others further away were told to remain indoors for several hours. Also, non-essential workers were sent home early. The surface soil sank two to four feet over a 400 square foot area. About 550 cubic yards of soil was used to fill the hole. There are a series of very long tunnels presumably containing large amounts of buried radioactive separation plant waste. The cause of the tunnel collapse was not revealed.[100]

Reactor Basin Sludges from 100 Reactor Area

The K-West and K-East Reactor basin sludges have been of great concern during the cleanup. These remaining sludges were highly radioactive. By 2019, the K-West reactor basin sludges had been removed to a "safe" location on top of the central plateau. The K-East reactor basin contained 20 feet of liquid overlaying the bottom sludges representing particles that had come off the reactor. The plan was to move the East basin sludges to the West basin, then place them in containers for transport and staging near the T- separations plant for interim storage. Twenty sludge containers had been gathered over a period of 15 months.[93,96,100]

The Tri Party Agreement of 1989 for Hanford Waste Cleanup– DOE, Washington State and EPA, Also, 2014 Act Establishing National Nuclear Parks

In 1989, DOE signed the Tri-Party Agreement requiring cleanup of the Hanford facility. DOE estimated the cleanup would take 30 years and cost $50 billion. Today these estimates are extremely low. The first hard decision was what to do with the 9 reactors lining the Columbia River. In 1992, DOE decided to "cocoon" the reactors by enclosing them in concrete walls and covered by metal roofs for over 75 years, after which they would be torn down and buried. B Reactor was last in this line of action because it would eventually be declared a national historic site and museum. In 2003, legislation was passed by interested parties that directed the National Park Service to conduct a feasibility study for "protecting" Manhattan Project sites. The document was revised to create a fully managed park not only at Hanford but also Los Alamos and Oak Ridge. In 2014, President Obama signed the National Defense Authorization Act with provisions to establish a National (Nuclear) Park.

The Tri-Party Agreement led to the creation of a Hanford Advisory Board in 1995 to provide cleanup guidance for the three agencies involved. These agencies shared regulatory oversight based on Hanford being a Superfund and RCRA site. Hanford represents the world's largest environmental cleanup. The Board's greatest concern today is the amount of work that remains to clean up Hanford, despite the huge sums that have already been spent. The leaders of the Manhattan Project did not devote much thought to the mess they were creating. Now the bill became due, and it was immense. DOE has estimated cleaning up Hanford would cost at least $300 billion and possibly more than $600 billion, much more than the cost of establishing and operating Hanford over its history. It seems unlikely the federal government will provide the required funding to complete cleanup. At Hanford, six of its nine reactors have been cocooned, and two will follow.[1,93,100,101]

Congressional Committee Questions DOE in Escalating Cleanup Delays and Liabilities

In May 2019, a Congressional Committee met to review DOE's management and cleanup of various nuclear sites. It said 16 sites including Hanford continued to constitute a massive effort to achieve cleanup. These 16 nuclear sites are shown in Figure 12.6. Government environmental liability for nuclear plants was in 2019 at $577 billion and still rising, with a final date for completion between 2070 and 2075. One report said cleanup could last until 2080. DOE was told to get its act together. GAO and the government said a culture change was necessary in the DOE and they were unhappy with the cleanup. A second report said complete restoration across the DOE network was estimated to have a cost of $500 billion to $1 trillion, with Hanford being the most expensive. It was said every element in the periodic table could be found in the soils and waters of nuclear materials production sites around the country. Some experts and lawmakers were becoming skeptical that a complete cleanup of the nation's weapons facilities was even possible and that there was not enough money in the world to clean all of it up. This has not been helped by decreasing budget allocations and apparent declining incentive in recent times to get the job done.[1,11,12,93,98]

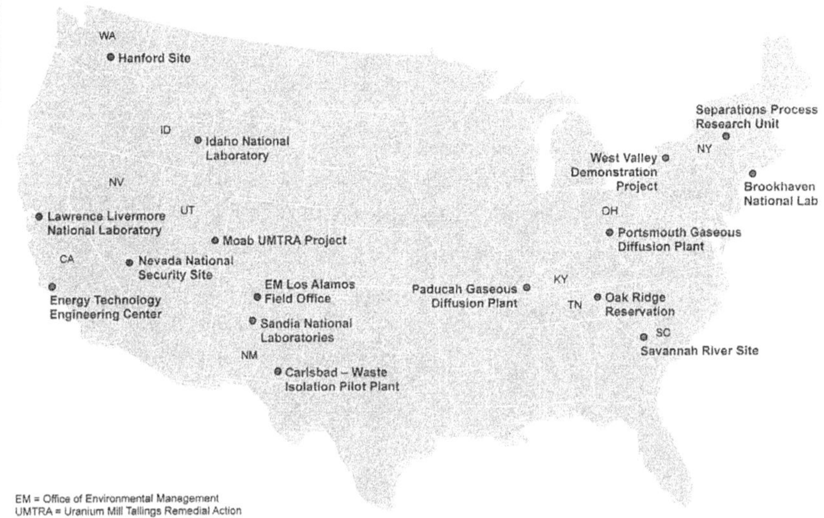

Figure 12.6 Nuclear Weapons Sites Across U.S. Requiring Waste Cleanup, May 2019

Credit: HR Subcommittee Report, USDOE and YouTube

Ongoing Hanford Waste Cleanup is Difficult, Behind Schedule and Very Costly

Old dumps have been excavated together with contaminated soils and the wastes moved to a huge landfill near the separation plants. One may ask if this is a satisfactory solution. A large plume of underground wastewater was reported below the tank farms at the separation plants. Gigantic pump and treatment stations across the Hanford site now lift up the groundwater, remove contaminants that go to landfills, and reinject the treated water back into the ground. Adjacent aquifers still contain an estimated 270 billion gallons of contaminated groundwater. Apparently, these extraction and treatment facilities have slowed down the flow of radioactive and toxic wastes going toward the Columbia River. There are five pump and treat facilities for dealing with contaminated groundwater. Around two billion gallons of contaminated groundwater are being treated annually, and a total of 27 billion gallons have been treated to date, which are returned to the ground. These groundwater operations are shown in Figure 12.7.[1,11,93,97]

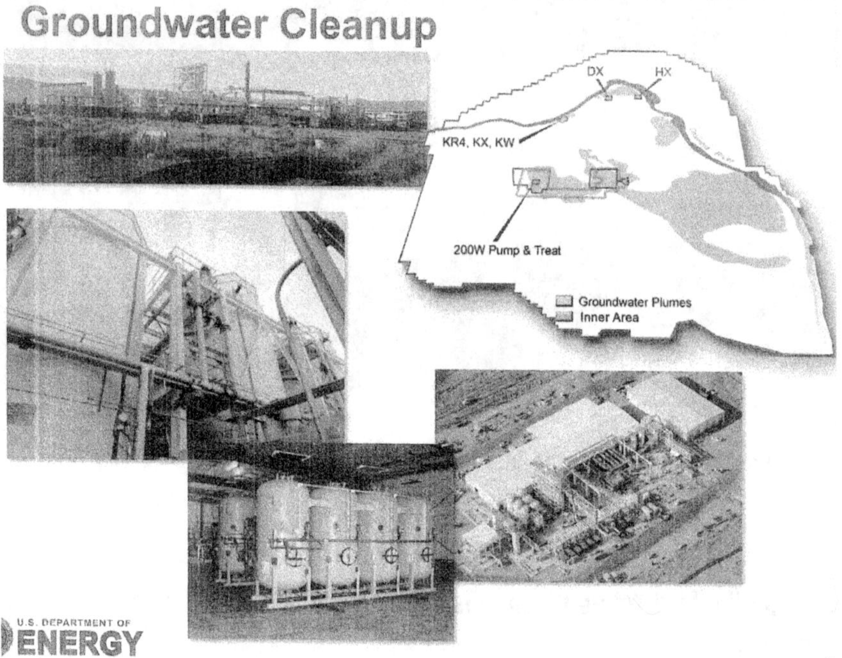

Figure 12.7 Hanford Groundwater Cleanup Operations, May 2013
Credit: USDOE, Hanford Site Overview

The most difficult and expensive cleanup tasks are just beginning. The high-level radioactive wastes continue to sit in the single-shelled and double-shelled tanks around the canyon buildings. The plan is to vitrify these wastes and place them in a high-level waste repository, but the process is expensive, difficult to achieve, and has barely started. Figure 12.8 shows part of the low level waste vitrification container handling system. Presumably, the

K-West and K-East reactor sludges will eventually be designated for vitrification. There will be high-level and low-level vitrification plants. The expected product will be "glassified logs". The vitrification product even if successful, has nowhere to go since the U.S. has not designated a repository for its high-level military and civilian nuclear wastes. Figure 12.9 shows some of the large high level waste storage tanks on site.[1,11,93,97]

Figure 12.8 Hanford Facility, Low–Level Waste Vitrification
Credit: USDOE and Bechtel

Figure 12.9 Hanford Facility, Large Waste Storage Tanks
Credit: USDOE and Bechtel

A program update was given in mid-2021 for the Hanford facility.[95] There seemed to be some questions among Tri Party Agreement members as what constitutes or defines high-level radiological wastes, and the specific means of safe disposal. Hanford was carrying an annual budget of $2.5 billion with 1,100 employees and six contractors involved in cleanup and facility upkeep. A start has been made on how to deal with low-level radioactive waste from the K-East and K-West reactor sludge holding basins. The single-shell tanks in the tank farms represent a substantial problem, and the vitrification treatment plant for handling these wastes will not start until 2023. The last two K Reactors will soon be cocooned. As regards to what is in the tank farm wastes, Hanford said … just look at the periodic table. Cesium and strontium are very important. Hanford provided a summary of remediation to date, but did not indicate what is remaining:

"Reactors- 6 cocooned and 1 preserved.
922 buildings or facilities demolished.
1,354 waste sites remediated- although another report said 1,800 sites.
12,687 cubic meters of plutonium contaminated waste retrieved.
18.7 million tons of solid debris moved to landfill.
100% of spent fuel moved to storage.
27 billion gallons of groundwater treated.
674 tons of contamination removed from groundwater."[95,98]

Nuclear Museum and National Nuclear Historic Site

The author Steve Olson in 2020 said the B Reactor Museum is not easy to reach and sits about a half mile from the Columbia River. Unfortunately, it fails to recognize the Hiroshima and Nagasaki events. The U.S. and Russia possess more than 90% of the world's nuclear weapons but both have refused to adopt nuclear weapons termination as an objective. Americans continue to spend an average of hundreds of dollars per person per year to maintain and upgrade their nuclear weapons. If future crises have at least the possibility of going nuclear, human civilization may not survive. Despite brilliance of its technological achievements, the story of Hanford has been a story of human misunderstanding.[1]

EPILOGUE

The Office of Technology Assessment (OTA) Report of February 1991[12] offers us a picture of where the nuclear weapons industry was some 30 years ago and can serve as a guide as what progress has been made environmentally since- or perhaps not, if their recommendations have not been heard. Progress is clouded because information is not easy to obtain. OTA focused on health exposure and necessary waste cleanup.

OTA said that "DOE has historically avoided public notification of releases from the weapons plants and their possible health effects. This practice has created substantial public distrust of DOE's methods and motivation." DOE has consistently maintained that no current contamination situations pose an "imminent threat" to public health. Information about off-site contamination had been limited, especially the potential for offsite human exposure. DOE's assertion there are no imminent health risks may be correct, but this had not been substantiated by scientific evidence. Immediate and long-range health impacts to workers and the public requires much more extensive evaluation including chronic health effects. Urgent health-based remediation requires developing environmentally and human-safe health-based cleanup priorities which were not supported by the prevailing regulatory processes.

A consensus gathered by OTA in the early-1990s was that cleanup costs would be in the tens of billions of dollars if proper procedures were followed- but today's total cumulative costs seem to be in the trillions of dollars. No consensus has been reached on how and where to dispose of these wastes. Much of the waste generated in the past and that generated in the future is destined to remain where it was created, and for decades to come. OTA said cleanup is troubled by three main problems: 1) lack of technical and institutional resources and processes; 2) lack of DOE past creditability, full public disclosure, and mechanisms for involving the public in the cleanup; and 3) lack of a comprehensive strategy for evaluating potential offsite human exposure and the overall health impacts of contamination. Recommended policy initiatives included: more Congressional oversight, stronger agency personnel, plan for safe waste storage, more technology development, more public access to information, coordination and acceleration of standard-setting, and better site monitoring programs.[12] This author believes that far from adequate advances have been made in these areas over the many past decades.

The OTA in 1991 advised that the public be provided with all relevant information on waste management and environmental situations. In spite of DOE's declassification efforts in 1993-1994, the process should further be accelerated, that the public promptly receive the requested material; and they be given notification of meetings, hearings, comment periods, etc. Remediation methods may have limited capabilities. Therefore, many sites may never be returned to a condition suitable for unrestricted public access. It may be nearly impossible to remove all the contaminants from groundwater and deeply buried soil, these methods are too expensive, require an extraordinary long time period, or necessary technologies are not yet available.[12]

The human species always has had the need to bury its best and worse toys and fears underground. With climate change and plastics, we also exploit the resources of the sea. We almost unconsciously bend to the will of convenience, politics and big money. Are AI, the i-phone and the Internet taking away our ability to self-understand? Are we seeing the undercurrents of defense and terrorism? The author believes that nuclear power is perhaps our greatest threat because it can be so quick, efficient and horrifying in destroying our planet. Is it too late to turn things around? We shall take a very brief look at SMRs or Small Modular Reactors which have become popular, the extreme wartime situation in Ukraine with looming nuclear threats and possible catastrophic damage to the largest commercial reactor in Europe- the Zaporizhzhia nuclear reactor, and the present status of nuclear reactors and nuclear power.

Small Nuclear Power Reactors

The World Nuclear Association (WNA) has been emphasizing "small and simpler units" for generating electricity from nuclear power and for process heat. Increased interest in "small" and medium nuclear power reactors is being driven by the desire to reduce capital costs and provide power for places distant from large grid systems. Economies of scale are said possible due to the large numbers of units expected to be manufactured. The International Atomic Energy Agency (IAEA) defines "small" as being under 300 MWe, and medium up to around 700 MWe. A subcategory of very small reactors is also proposed for units less than 15 MWe, especially for small remote communities. The modular concept is utilized, and the units are assembled offsite. Four main options include: light water reactors; fast neutron reactors; graphite- moderated high temperature reactors; and molten salt reactors. Nevertheless, a number of challenges remain, including performance, safety and licensing among other challenges. The WNA report cites 5 Small Modular Reactors (SMRs) in operation ranging from 11 to 300 MWe, by developers in Pakistan, China, and Russia. Four SMRs are under construction with capacities ranging from 27 to 300 MWe in Argentina, China and Russia.[101]

In January 2022, the EU added nuclear power to its list of projects now eligible for "green" financing. The nuclear industry is claiming a new generation of small nuclear reactors (SMRs) designed to be cheaper, quicker and less financially risky to build. China hopes to see its first commercial SMR operating in Hainan by 2026. NuScale Power, an American firm, hopes its first SMR, being built at the Idaho National Laboratory, will provide power by 2029. But NuScale's Idaho plant saw its costs increase from $3.6 billion in 2017 to $6.1 billion in 2020, and several of its commercial partners pulled out of the project in 2020. Small scale predicts higher construction costs, which may not be overcome by mass production practices.[102] The U.S. is moving to help Romania develop a SMR. This new plant style is a smaller, less costly modular reactor, which still appears controversial. The U.S. said it would supply Romania with a training simulator to prepare for the building of a SMR in Romania by NuScale to be ready by the end of the decade. The Romanian project envisions a power station comprised of six modular units generating 462 MWe and costing around $1.6 billion. The simulator would be for training Romanians to operate one of the reactors and would be paid for by the U.S.[103]

The Zaporizhzhia, Ukraine Nuclear Reactor

The International Atomic Energy- the UN nuclear watchdog, reports that intense shelling of this nuclear complex in southern Ukraine taken over by Russia in March 2022, has caused widespread damage across the site, but that key equipment has not yet been affected. The head of the agency further said a major incident had been avoided by "meters" and not by "kilometers."[104] Rafael Marino Grossi, the head of the UN nuclear agency said every principle of nuclear safety has been violated at Europe's largest nuclear plant occupied by Russian troops starting in February 2022. The situation was out of control, and extremely grave and dangerous. Fighting around the plant caused a fire that led to global concerns about a possible nuclear accident. Since March 2022, Russia has greatly built up its military forces and war equipment within the Zaporizhzhia plant. It is believed that all reactors at Zaporizhzhia are currently offline, but cooling water needs, and other upkeep must be maintained. The Ukrainian operational staff has been decimated. What further destruction might be made to the power plant by the Russian military, if and when they leave the plant?[105]

Past and Future Status of Nuclear Power

Serhii Plokhy in his book "Atoms and Ashes" describes six major nuclear accidents in the past- both military and civil- that have plagued the nuclear industry.[4] He relates a historic vote at the Nuclear Regulatory Commission (NRC) headquarters in North Bethesda, Maryland on February 9, 2012, where two new nuclear reactors were approved, the first licensed after the March 1979 Three Mile Island accident. Gregory Jaczko, the Chairman of the NRC was outvoted four to one by his own Commission, and he stepped down as Chairman in May 2012. He was vocal saying he could not support this license as if Fukushima never happened, and if the Southern Company, that made application for the license, did not make a binding commitment for necessary enhancements rejected by his fellow commissioners (a familiar past pattern). The CEO for the Southern Company suggested these enhancements were not applicable to this newest nuclear technology. The original estimated cost of the new reactors was $14 billion but had gone up to $25 billion in 2022 with delays. The project would be largely supported by the federal government. The World Nuclear Association predicts an enormous increase in the number of nuclear reactors.[4,101]

Jaczko was not the only one predicting accidents, and further said… "as the pressure to cut costs increases, safety will suffer", and profit will control. Accidents will happen and more are inevitable. Most of these are predictable and can be dealt with. Others get out of hand for reasons of technology or because of human factors, causing damage and sometimes resulting in the release of radioactivity into the atmosphere. And often, government monitors and regulators established a friendly relationship with industry, dismissing safety violations.[4]

With many old risk factors remaining and the emergence of new ones, it is difficult to be optimistic about a future free of nuclear accidents. An unresolved issue is the design of reactors, which came from military prototypes for producing plutonium or for powering nuclear submarines. Another major problem is the disposal of spent fuel, for which there is no easy solution to be hopefully solved in the future. A rapid expansion of nuclear power plants, proposed for dealing with climate change, will increase the probability of accidents. Bill Gates' claim that for the new generation of reactors, that… "accidents would literally be prevented by the laws of physics", can be questioned. It is very costly and takes too much time to build a reactor, and it is inherently unsafe over the long range not only for technological reasons, but also because of the risk of attendant error and accidents. Investing money in nuclear energy today means less attention paid to the expansion of renewables. For existing nuclear facilities, it is necessary to strengthen facility oversight, increase the safety and security of aging nuclear power plants, and invest suitable resources.[4]

A New York Times reporter in September 2022 attended the annual World Nuclear Symposium sponsored by the nuclear industry's global trade group. Overheard conversation was that the public has been brainwashed about the danger of nuclear power, but the industry's optimism could fall short in practice. Nuclear is far slower to build than most other forms of power, and it is far more expensive. A third problem is that as battery technology improves and the price of electricity storage drops, nuclear may be way too late, with much of its value eclipsed by cheaper, faster, and more flexible renewable power technologies.[106]

The 63 nuclear reactors that went into service between 2011 and 2020 took an average of about 10 years to build. Solar and wind farms can be built in months. In 2020 and 2021, the world added 464 gigawatts of wind and solar power generation capacity, which is said to be more power that can be generated by all the nuclear plants operating in the world today. The nuclear industry has been notorious for its cost overruns and delays. One new U.S. reactor came online in 2016 for the first time in 20 years. The only nuclear reactor under construction in the U.S. is at the Plant Vogtle power station in Georgia which was started in 2013 and projected to be finished in 2017, but which is not yet done. The initial budget for the facility was $14 billion, which has doubled to over $28 billion. In 2017, two reactors midway through construction in South Carolina were cancelled after the cost projections increased from \$11.5 billion to more than \$25 billion. Also, extraneous costs such as disasters are not included therein.[106]

Mark Jacobson of Stanford University said nuclear no longer makes sense because any new money put into nuclear is money not being spent on renewable projects that could lower greenhouse emissions much faster. The price of lithium-ion batteries has dropped by 97% since they were introduced in 1991. Jacobson is one of many who argue that such advances will render nuclear power essentially obsolete. The IEA says that nuclear capacity will need to double by 2050, especially in developing countries. But they also say by 2050, nuclear power will account for less than 10% of global energy compared to 90% for renewables. Small advanced reactors may become viable. However, it seems unlikely that nuclear can play close to a dominant role in the future, and it is plagued by problems.[106]

China is planning on at least 150 nuclear reactors in the next 15 years, more than what the rest of the world has built in the past 35 years. China has emerged as the world's last great believer in nuclear power plants. It has plans to build a large number of nuclear power plants at a relatively low cost. In 2021, the China General Nuclear Power Corporation said its goal was to add by 2035 some 200 GW of power, enough to power more than a dozen cities the size of Beijing. China says its plans will reduce about 1.5 billion tons of annual carbon emissions.[107]

The outcome of the Manhattan Project was to produce rockets principally for carrying nuclear weapons. Throughout the Cold War, we lived under the threat of nuclear catastrophe that could have brought down human civilization, a threat most acute at the time of the Cuban crisis in 1962. Robert McNamara spoke of this situation only after he retired. In his confessional documentary film "The Fog of War" he said… "We came within a hair's breadth of nuclear war without realizing it. It's no credit to us that we escaped-Khrushchev and Kennedy were lucky as well as wise." On several occasions during the Cold War the superpowers skirted Armageddon, and this prospect has not gone away. We cannot rule out that by mid-century there will be a standoff between new superpowers that will be handled less well than was the Cuban crisis. The risk that smaller nuclear arsenals proliferate and are used in a regional context, is higher than it ever was. Terrorists may one day acquire a nuclear weapon and willingly detonate it in a city, killing tens of thousands along with themselves. The past decision to spend enormous resources on instruments of destruction has been an immense historical tragedy. The nuclear age started an era where humans can threaten earth's future existence. We will never be completely free of the nuclear threat. H-bombs cannot be dis-invented, and we are subject to grave new perils.[1,108]

NOTES

1. The Apocalypse Factory: Plutonium and the Making of the Atomic Age, Steve Olson, 2020

2. A World Destroyed: The Atomic Bomb and the Grand Alliance, Martin Sherwin, 1975

3. The Manhattan Project, Wikipedia, 2021.

4. Atoms and Ashes: A Global History of Nuclear Disasters, Serhii Plokhy, 2022

5. African Americans Against the Bomb: Nuclear Weapons, Colonialism and the Black Freedom Movement, J. Intondi, 2015

6. Restricted Data: The History of Nuclear Secrecy in the United States, Alex Wallerstein, 2021

7. History of the Rocky Flats Plant: Nuclear Weapons Production Facility, Colorado, Atomic Bomb Components, Plutonium Contamination Issues, Environmental and Safety Concerns, Shutdown and Cleanup, U. S. Military, Department of Defense and Progressive Management Publications, November 2017.

8. The Ambushed Grand Jury: How the Justice Department Covered Up Government Nuclear Crimes and We Caught Them Red Handed, Wes McKinley and Caron Belkany, 2004.

9. Article on website of Rocky Flats Stewardship Council, September 1, 2021.

10. Secrets of the Atomic Bomb, YouTube, February 21, 2019

11. Secret Mesa: Inside Los Alamos National Laboratory, Jo Anne Shroyer, 1998.

12. Complex Cleanup: The Environmental Legacy of Nuclear Weapons Production, OTA-0-484, Congress of the U.S., Office of Technology Assessment, February 1991

13. DOE's Mounting Cleanup Costs: Billions in Environmental Liability and Growing, Subcommittee on Oversight and Investigations, Committee on Energy & Commerce, U.S. House of Representatives Committee Repository & YouTube, May 1, 2019.

14. The Paradoxes of Deterrence: How the Debate About Nuclear Weapons Has Evolved, Commonweal, 3/2021.

15. Report- Disposition and Control of Uranium Mill Tailings Piles in the Colorado River Basin, USDHEW, FWPCA, Region 8, Colorado River Basin Project, March 1966.

16. The Legacy of the Manhattan Project and Definition of Curie, YouTube, March 11, 2013.

17. U.S. Army Corps of Engineers website.

18. Wikipedia, curies, September 27, 2021

19. Full Body Burden: Growing Up in the Shadow of Rocky Flats, Kristen Iversen, 2013

20. Uranium Location Database Compilation, EPA Office of Radiation Protection Division, Washington, D.C., EPA 402-R-05-009, August 2006.

21. About Abandoned Uranium Mines, BLM Web Site, post-1998.

22. The True Impact of Uranium Mining, YouTube, July 18, 2016.

23. How is Uranium Mining Conducted in the U.S., YouTube, July 10, 2012.

24. Uranium Mining and Milling Wastes, Peter Diehl, World Information Service on Energy (WISE), last update of May 19, 2011.

25. Mining, Milling, Conversion and Enrichment of Uranium Ores, YouTube, Oak Ridge, November 17, 2021.

26. Article- Trump's $1.5 Billion Uranium Bailout Triggers Rush of Mining Plans, U.S. News and World Report, February 14, 2020.

27. Article- Uranium Firm urged Trump Officials to Shrink Bear Ears National Monument, Washington Post, December 8, 2017.

28. The Political Economy of Radioactive Colonialism, Ward Churchill and Winona LaDuke, Chapter 8 from the book "The State of North America", M. Annette Jaimes Ed., 1992.

29. Article- Tapped Out, Time Magazine, March 2, 2020.

30. How the U.S. Poisoned Navajo Nation, YouTube, October 21, 2020.

31. Native Perspectives on Uranium and the Environmental Justice #1, YouTube, Grand Canyon Trust, November 23, 2010.

32. First Session of the Conference in the Matter of Pollution of the Interstate Waters of the Colorado River and Its Tributaries, USDHEW, USPHS January 13, 1960, Phoenix, Arizona.

33. Second Session of the Conference in the Matter of Pollution of the Interstate Waters of the Colorado River and Its Tributaries, USDHEW, USPHS, May 11, 1962, Las Vegas, Nevada.

34. Third Session of the Conference in the Matter of Pollution of the Interstate Waters of the Colorado River and Its Tributaries, USDHEW, USPHS, May 9-10, 1962, Salt Lake City, Utah.

35. Fourth Session of the Conference in the Matter of Pollution of the Interstate Waters of the Colorado River and Its Tributaries, USDHEW, USPHS, May 27, 1963, San Diego, California.

36. Interstate Conference on Pollution of the Colorado River and Its Tributaries, February 13-15, 1963, Santa Fe, New Mexico, USDHEW, USPHS, Preparatory Report prior to the Technical Conference.

37. Fifth Session of the Conference in the Matter of Pollution of the Interstate Waters of the Colorado River and Its Tributaries, USDHEW, USPHS, May 26, 1964, Las Vegas, Nevada.

38. Sixth Session of the Conference in the Matter of Pollution of the Interstate Waters of the Colorado River and Its tributaries, USDHEW, USPHS, July 26, 1967, Denver, Colorado.

39. Seventh Session of the Conference in the Matter of Pollution of the Interstate Waters of the Colorado River and Its Tributaries, USEPA, February 15-17, 1972, Las Vegas, Nevada.

40. Proposed Model for Inactive Tailings Pile Regulation, USEPA, Region 8, February 1972.

41. Dear Sir: Your House is Built on Radioactive Uranium Waste, Peter Metzger, New York Times Archival Service, October 31, 1971.

42. Summary Report, Phase I Study of Inactive Uranium Mill Sites and Tailings Piles, October 1974. This report was in response to the Subcommittee on Raw Materials of the Joint Committee on Atomic Energy held re: S. 2566 and H. 11378 of March 12, 1974.

43. Guidelines for Cleanup of uranium Tailings from Inactive Mills, DOE, Office of Scientific and Technical Information (OSTI), January 1, 1975.

44. Uranium Mill Tailings Radiation Control Act, PL 95-604, 95th Congress, aka Uranium Mill Tailings Radiation Control Act of 1978 or UMTRCA, November 8, 1978.

45. GAO Report to Congressional Committee on Uranium Mill Tailings: Cleanup Continues but Future Costs Are Uncertain, Executive Summary, December 1995.

46. UMTRCA Title I and II Fact Sheet: Uranium Mill Tailings Radiation Control Act Sites, DOE Legacy Management, October 24, 2014.

47. Toxic Legacy of Uranium Haunts Proposed Colorado Mill (Grand Junction CO and other mills), Nancy Lofholm, Denver Post, May 5, 2016 update.

48. Article- Colorado Denies License for Paradox Uranium Mill, Durango Herald, April 30, 2018.

49. Uranium Mill Tailings Management Plan: For Managing Title I Uranium Mill Tailings Encountered During Construction Activities in Western Colorado, Colorado Department of Health and Environment, Updated June 2019.

50. Durango, CO. Uranium Mill Tailings Removal Collection (VCA Mill). Center of Southwest Studies, Fort Lewis College, 1990.

51. River of Lost Souls, Jonathan Thompson, 2018.

52. More Than 100 Properties Missed (at Durango, CO), Denver Post, August 4, 2019.

53. Durango, CO Disposal/Processing Site, DOE/LM Video, March 25, 2020.

54. Cleanup of Historic Uranium Mill Completed, News release by USEPA, September 29, 2008.

55. Uravan Uranium Project, Colorado Department of Public Health and Environment, ~2013.

56. Decommissioning of Moab, Utah Uranium Mill Tailings (Archive 1996-1999), last update of June 22, 2002, WISE Uranium Project.

57. Executive Summary of Critical Issues, Topic- Moab Tailings Pile, DOE, August 1, 2007.

58. Cleaning Up Mill Tailings and Groundwater at the Moab UMTRCA Project Site, DOE, August 2, 2010.

59. Moab Uranium Tailings Cleanup Still Going After 13 Years, YouTube, May 12, 2021.

60. Radioactive Waste Shipments (Alternative Fuel Materials) Shipped to White Mesa Mill, 1993-2016, Grand Canyon Trust, March 31, 2020.

61. Letter of January 8, 2020, Ute Mountain tribe to Utah Department of Environmental Quality, Salt Lake City re: Aquifer Pollution Around the White Mesa Uranium Mill.

62. Issues at White Mesa Uranium Mill (Utah), WISE Uranium Project, last update June 2, 2020.

63. Case Study 2- The White Mesa Uranium Mill in San Juan County, Utah, WISE Uranium Project, Issues at White Mesa Uranium Mill, August 1997 to Present, last update June 2, 2020.

64. The Business of Radioactive Waste, Grand Canyon Trust, March 15, 2020.

65. Portsmouth Gaseous Diffusion Plant, Wikipedia, September 18, 2021.

66. National Enrichment Facility, Wikipedia, September 18, 2021.

67. Uranium Enrichment- What is Driving Prices Higher? www.Prweb.com/releases.

68. OLLI Denver University Course, Great Rivers of the World, Spring 2021.

69. From Putin Country: A Journey into the Real Russia, Anne Garrels, 2016.

70. Plutopia: Nuclear Facilities, Atomic Cities, and the Great Soviet and American Plutonium Disasters", Kate Brown, 2013.

71. Ozyorsk Chelyabinsk Oblast, Wikipedia, April 24, 2021.

72. Colorado Experience: Colorado's Cold War: Rocky Flats, YouTube, September 17, 2014.

73. The Most Dangerous Building in America: Rocky Flats, YouTube & Channel 7 News, February 17, 2017.

74. Wikipedia article on Pits (Nuclear Weapons), August 11, 2021.

75. Nuclear Weapons Documentary: Rocky Flats Secrets of a Bomb Factory, YouTube and Frontline, March 12, 2018.

76. Colorado Department of Public Health and Environment, Summary of Findings, Historical Public Exposure Studies on Rocky Flats, August 1999.

77. Transuranic Waste- 2019, Idaho Leadership in Nuclear Energy Commission 3.0.

78. Buried Waste Retrieval Resumes at Idaho Site, DOE Office of Environmental Management, September 11, 2018.

79. U.S. Close to Ending Buried Nuke Waste Cleanup in Idaho, Denver Post, January 10, 2022.

80. Opposition Remains to Opening Wildlife Refuge for Public Use (Rocky Flats), Denver Post, May 16, 2017.

81. At Work in the Atomic City: A Labor and Social History of Oak Ridge, Tennessee, R. Olwell, 2004.

82. Oak Ridge National Laboratory, Wikipedia, May 20, 2021.

83. The Story of Mercury Use and Environmental Contamination at the Oak Ridge Y-12 Plant, ORNL, January 2010.

84. Los Alamos National Lab, Wikipedia, May 5, 2021.

85. CDC, Radiation Emergencies, April 4, 2018.

86. Los Alamos Will Never Be Clean: Trinity 70 Years Later, The New Mexican, July 13, 2015.

87. Bayo Canyon Aggregate Area, New Mexico Site, YouTube, August 22, 2020.

88. Environmental Groups: Los Alamos County Water Pollution is Public Health Threat, Amigos Bravos, Western Environmental Law Center, June 28, 2019.

89. First Amendment Audit- Pantex Nuclear Manufacturing Plant, YouTube, October 4, 2017.

90. Pantex, Wikipedia, July 5, 2021.

91. Why These Old Weapons Never Die, New York Times, November 22, 2022

92. Savannah River Site, Wikipedia, July 4, 2021.

93. The Hanford Site, Wikipedia, May 5, 2021.

94. Michele Gerber's Interview, YouTube, April 12, 2019.

95. Hanford Live 2021, YouTube, 6/16/2021 and Tri-City Herald, Annette Cary, May 1, 2019.

96. Hanford Sludge Removal Success, YouTube, October 1, 2019.

97. The Hanford Story: Tank Waste Cleanup, YouTube, December 1, 2016.

98. The Area: A Journey Through the Hanford Nuclear Reservation, YouTube, October 18, 2011.

99. Video Shows Illegal Dumping of Toxic Liquids at Hanford, www.kgw.com., October 27, 2017.

100. Tunnel at Nuclear Waste Storage Facility Collapses (Hanford Reservation), Denver Post, May 10, 2017.

101. Small Nuclear Power Reactors, World Nuclear Association, Updated May 2022.

102. Nuclear Energy: Pint-Sized Power Stations, The Economist, March 26, 2022.

103. U.S. Moves to Help Romania on New Nuclear Plant Style, New York Times, May 24, 2022.

104. Nuclear Energy: Pint-Sized Power Stations, The Economist, November 25, 2022.

105. UN Nuclear Chief Warns of Extremely Grave Risk at Power Plant, New York Times, August 5, 2022.

106. Nuclear Power Still Doesn't Make Much Sense, Farhad Manjoo, New York Times, September 17, 2022.

107. China's Climate Goals Hinge on a $440 Billion Buildout, Bloomberg Green, November 6, 2021.

108. From Here to Eternity: A Vision for the Future of Science, Martin Rees, 2012

Additional References

109. Message from Mr. O'Halleran, Navajos in Arizona to M. Wheeler, EPA re: abandoned mines on Navajo lands.

110. Moab Uranium Tailings Cleanup Still Going After 13 Years, YouTube, May 12, 2021.

111. The Three Mile Island Nuclear Accident, New York Academy of Sciences, Vol. 365, 1981.

112. Small Nuclear Reactors, IAEA, 2020.

113. Voices from Chernobyl: The Oral History of a Nuclear Disaster, S. Alexievich (Translated), 2005.

114. Still A Radioactive No-Go Zone (Fukushima, Japan), Denver Post, March 11, 2021.

115. Still Recovering, Japan Marks 10 Years Since Tsunami, Denver Gazette, March 12, 2021.

116. Japan: Can It Move Past Fukushima's Toxic Legacy, The Week, March 19, 2021.

117. Materials from Colorado: A Historical Atlas, Noel, 2015.

118. Ozyorsk Chelyabinsk Oblast, Wikipedia, April 24, 2021.

119. Three Mile Island Accident, Wikipedia, May 2, 2021.

120. Durango, CO Disposal/Processing Site, YouTube, April 24, 2020.

121. Three Mile Documentary, YouTube, 2015.

122. Colorado Experience: Uranium Mania, YouTube, November 2, 2017.
123. The Discrete Charm of Nuclear Power, The Economist, November 13, 2021.
124. Bill Gates' Nuclear Startup Picks a Wyoming Town for Its First Advanced Reactor Which Will Cost $4 Billion, Microsoft News, November 18, 2021.

ABOUT THE AUTHOR

Born on Long Island, New York, Ed Struzeski Jr. attended Manhattan College for his BCE and then received his MSCE from North Carolina State University. He underwent additional studies at the University of Cincinnati, University of Colorado and Denver University. He has been in Denver, Colorado for more than 50 years and has traveled to all 50 states and many countries of the world. He is a firm believer in the American West especially with its skiing, hiking and climbing of its mountains, plains and deserts. Ed is an environmental, chemical, civil engineer with more than 50 years in the environmental field. He was a Commissioned Officer in the U.S. Public Health Service serving with different Federal agencies including the USEPA, and in the last 10 working years in consulting engineering. In the early stages of his career, he was involved in the study of uranium mines and mills mainly in the seven-state region of the Colorado River Basin, and the impact of these and other sources on the radiological and chemical properties of rivers and waterways in the Basin. His special professional areas were in river water quality studies, environmental audits, and domestic and industrial process wastewaters and their control and treatment. Ed was a member of the FBI-EPA team for the "sting" raid on the Rocky Flats nuclear facility on June 6, 1989, but did not participate in the actual event due to pending EPA retirement, the first of three retirements.